ANGULAR MOMENTUM CALCULUS IN QUANTUM PHYSICS

ANGULAR MOMENTUM CALCULUS IN QUANTUM PHYSICS

Michael DANOS
National Institute of Science and Technology
Gaithersburg, USA
and

Vincent GILLET
Départment de Physique Nucléaire
CEN-Saclay, Gif-sur-Yvette, France

World Scientific
Singapore • New Jersey • London • Hong Kong

Published by

World Scientific Publishing Co. Pte. Ltd.
P O Box 128, Farrer Road, Singapore 9128
USA office: 687 Hartwell Street, Teaneck, NJ 07666
UK office: 73 Lynton Mead, Totteridge, London N20 8DH

Library of Congress Cataloging-in-Publication data is available.

ANGULAR MOMENTUM CALCULUS IN QUANTUM PHYSICS

ISBN 981-02-0412-4

Printed in Singapore by JBW Printers & Binders Pte. Ltd.

TABLE OF CONTENTS

Part Two

APPLICATIONS

INTRODUCTION

To the reader

This book will teach you the skills needed to handle problems that both theoreticians and experimentalists may encounter in any angular momentum calculations of quantum physics. You will learn to treat these problems in the simplest possible way, i.e. in a framework which is the same for all cases, which will require a minimal use of formularies and will eliminate all phase difficulties. The method is simple to use, it is rapid and transparent, and it allows easy checking of the results.

Angular momentum theory itself is not the subject of this book. There exist many excellent texts on angular momentum in quantum theory. We aim instead at providing practical, efficient, and operative techniques which lead directly to solving actual problems. To use this manual the reader needs only a basic minimal knowledge of the concepts of angular momentum quantum mechanics, found in any of the familiar textbooks.

Angular momentum calculations are unwieldy in many situations : complicated operators, many-body systems, Fock space operators and state vectors, handling of anti-particles, reactions, polarizations, angular distributions, correlations, etc...The simplicity and power of the present method in treating these situations is based on the introduction and consistent use of invariants for all physical quantities : states, operators, experimental set-ups, etc... The convenience to deal only with invariants is possible and achieved by a judicious choice of the phase convention for the quantities which enter these invariants.

Everything else then is a general recoupling operation which is recognized to factorize into a succession of single basic unitary recoupling transformations. This is the key to the use of an extremely simple graphical representation for the complete evaluation of any problem.

The Invariant Graph method employed in the present book differs from previous graphical methods by its underlying concepts. Its rules and basic graph elements are self-evident and few. Any complete calculation is represented by a single graph which by direct reading immediately yields the final result, including all selection rules. The drawing of this graph is actually a lever for obtaining in the simplest, most economical, and physically informative way the algebraic form of the result.

All those who, like the authors of this book, have toiled over multiple interlocking intermediate sums, incomprehensible phases, and have fought with the sheer size and complexity of the geometrical factors of any somewhat involved process, will appreciate the logical transparency and the compactness of the present method.

Organisation of the book and how to use it.

Now a few words about the plan. In a first part the reader is directly presented with the Invariant Graph method. The second part then shows how the method is implemented in the most frequently encountered cases. In fact the reader who whishes to immediately start applying the method to his own problems may need only to read Chapters 1, 2 and 4 for acquiring the tools and to refer to Chapter 10 for a complete formulary.

In the first part, up to Chapter 4, we have adopted for pedagogical reasons a somewhat non-systematic but pragmatic approach. Our aim is to teach our reader by means of concrete examples, allowing him to put to work immediately the very few rules of the method. The reader will acquire an understanding and feel for the structure and the logic of the method in a practical step-by-step manner.

In the second part, Chapters 5 through 9, the Invariant Graph method is applied to essentially all the situations where angular momentum handling is central to the formal evaluation of the problem : matrix elements of differential operators, treatment of symmetrized many-body systems, calculations in Fock space, handling of the particle-hole representation, evaluation of transition matrix elements and angular correlations, density matrix descriptions. This second part will provide the reader with the basic expressions needed in these different problems, together with the complete steps of their calculation, however without going into the extraneous details which are specific of particular situations.

The summary of the method and the formularies needed for its implementation are given in Chapter 10, at the end of the book.

Definitions and phase convention

The Wigner coupling (3-j's) and recoupling (6-j's, 9-j's) coefficients and rotation matrices have in this work their usual definitions, as given for example in A.R. Edmonds, *Angular Momentum in Quantum Mechanics*, 1957 (Princeton University Press, Princeton N.J.), in L.C. Biedenharn and J.D. Louck, *Angular Momentum in Quantum Physics*, 1981 (Addison-Wesley P.C., Reading Mass.), or in U. Fano and G. Racah, *Irreducible Tensorial Sets*, 1959 (Academic Press, New York, N.Y.).

For convenience and easy reference we give here the principal notations and phase definitions used throughout, which will be explained in detail in the book.

All tensorial quantities $\Psi_M^{[I]}$ are expressed as *contrastandard* tensors (see Chapter 3), which is denoted by the square bracket superscript. We impose on all tensors the phase convention under hermitian conjugation,

$$\Psi_M^{[I]\dagger} = (-)^{I+M} \tilde{\Psi}_{-M}^{[I]}. \tag{I.1}$$

The tilda notation means that all matrices or spinors contained in the tensor undergo the transpose operation. For example for the spin wave function $s = ½$, $m = ½$, we have

$$\chi_m^{[s]} = \begin{pmatrix} 1 \\ 0 \end{pmatrix}, \qquad \chi_m^{[s]\dagger} = (1\ 0). \tag{I.2}$$

The hermitian conjugation for a half-integer spin wave function thus writes as a ket to bra transformation, defining the tilda operation on that spinor

$$\chi_m^{[s]\dagger} = (-)^{s+m} \tilde{\chi}_{-m}^{[s]}, \qquad \text{for example} \quad \tilde{\chi}_{-½}^{[s]} = (-1\ 0). \tag{I.3}$$

In order for orbital wave functions to obey this basic phase convention, we introduce the spherical harmonics $Y_m^{[\ell]}(\theta,\varphi)$ which are given in terms of the usual spherical harmonics $Y_{\ell m}(\theta,\varphi)$ by the relation

$$Y_m^{[\ell]}(\theta,\varphi) = (-i)^\ell\, Y_{\ell m}(\theta,\varphi), \tag{I.4}$$

with

$$Y^*_{\ell m}(\theta,\varphi) = (-)^m\, Y_{\ell -m}(\theta,\varphi). \tag{I.5}$$

These orbital wave functions indeed fulfill the phase convention (I.1) under hermitian conjugation

$$Y_m^{[\ell]}(\theta,\varphi)^\dagger = (-)^{\ell+m}\, \tilde{Y}_{-m}^{[\ell]}(\theta,\varphi) = (-)^{\ell+m} Y_{-m}^{[\ell]}(\theta,\varphi). \tag{I.6}$$

Here the transpose operation (denoted by the tilda symbol) has no effect. However the tilda notation will often be retained in the expressions for non-spinor quantities when it is important to distinguish whether these quantities originate from a bra or a ket.

In Fock space the tilda symbol also acquires the meaning of denoting the creation operators, since the hermitian conjugation of an annihilation operator $a_m^{[j]}$ results in a creation operator

$$a_m^{[j]+} = (-)^{j+m}\, \tilde{a}_{-m}^{[j]}. \tag{I.7}$$

The authors thank their colleagues and students who helped develop the examples given in this book. They gratefully acknowledge the precious support of Mrs. Th. Lecuyer who unflinchingly prepared and typed the manuscript.

Part one

THE GRAPHICAL METHOD

Chapter 1

RECOUPLING GRAPHS

This chapter contains the basic elements for drawing recoupling graphs. There is a one-to-one correspondence between every graph symbol and an algebraic expression. The required elements and rules are very few and center around a basic recoupling box involving four tensors. At this stage, upon having acquired this minimal information, it is already possible to use the graph method for deriving theorems on recoupling identities.

Coupling of tensors

The graphical representation of a tensor $\psi_M^{[I]}$ is a single horizontal line, fig.1.1.

1.1

The coupling of two tensors by means of Clebsch-Gordan vector coupling coefficients to total angular momentum I,M is denoted by brackets

$$[A^{[J]}B^{[K]}]_M^{[I]} = \sum_{m_J m_K} (JKm_J m_K|IM)A_{m_J}^{[J]}B_{m_K}^{[K]}. \tag{1.1}$$

It is graphically represented by two horizontal lines with a vertical coupling bracket, fig.1.2.

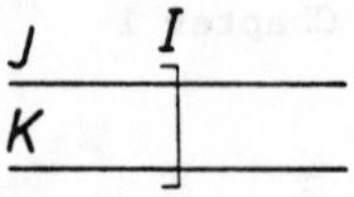

1.2

The total M value is omitted on the graph as will always be the case throughout the book. Where the orientation in space of the object will be physically significant the density matrix formalism will be used.

The coupling of three or more tensors follows the same notation. For example, the coupling of three tensors

$$T^{[I]}_{M} = [A^{[J]}[B^{[K]}C^{[L]}]^{[R]}]^{[I]}_{M}, \tag{1.2}$$

is graphically represented by fig.1.3.

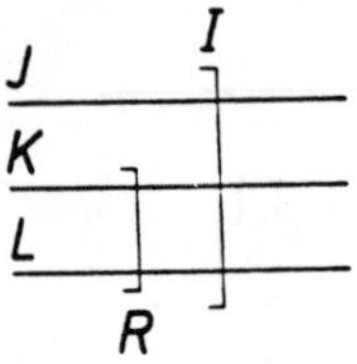

1.3

For four tensors

$$T^{[I]}_{M} = \left[[A^{[J]}B^{[K]}]^{[R]}\ [C^{[L]}D^{[P]}]^{[S]}\right]^{[I]}_{M}, \tag{1.3}$$

the graph is given by fig.1.4.

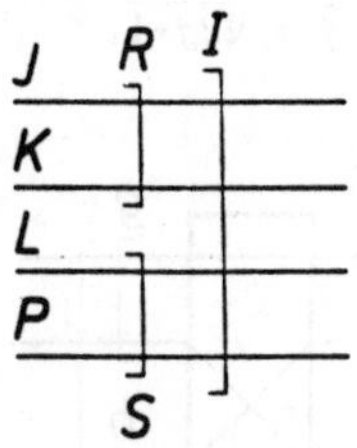

1.4

These graphical representations contain the complete information, i.e. they specify the individual tensorial objects, the intermediate and overall quantum numbers, and the coupling scheme.

The basic recoupling graph

We now develop the graphical representation of recoupling operations applied to coupled tensors.

All possible recoupling transformations can and will be performed using as the building block the basic recoupling transformation associated with four angular momenta

$$[[{}_{\psi}[a]{}_{\psi}[b]]^{[e]}[{}_{\psi}[c]{}_{\psi}[d]]^{[f]}]^{[i]} = \sum_{gh} \begin{bmatrix} a & b & e \\ c & d & f \\ g & h & i \end{bmatrix} [[{}_{\psi}[a]{}_{\psi}[c]]^{[g]}[{}_{\psi}[b]{}_{\psi}[d]]^{[h]}]^{[i]}. \tag{1.4}$$

It is represented by the graph of fig.1.5. We shall refer to the recoupling coefficients of this unitary transformation as square 9-j's since they are related to the Wigner 9-j coefficients by

$$\begin{bmatrix} a & b & e \\ c & d & f \\ g & h & i \end{bmatrix} = \hat{e}\,\hat{f}\,\hat{g}\,\hat{h} \begin{Bmatrix} a & b & e \\ c & d & f \\ g & h & i \end{Bmatrix}. \tag{1.5}$$

We shall always use the shorthand notation

$$\hat{j} = \sqrt{2j+1}. \tag{1.6}$$

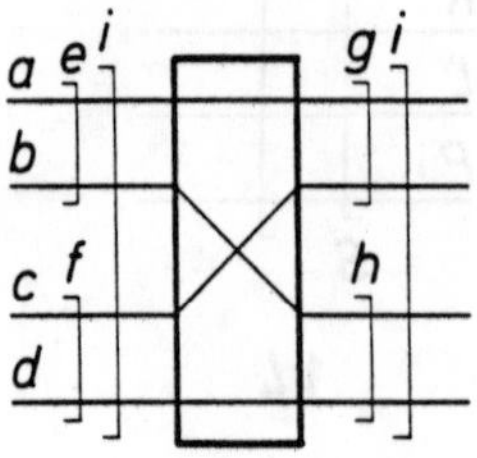

1.5

The rules followed in drawing this graph are :

(i) On the left side beginning at the top we draw lines associated with the initial coupling scheme, read from left to right. Each line is labelled by its angular momentum. Each coupling is represented by a bracket labelled by the corresponding angular momentum.

(ii) The recoupling coefficient is represented by a box with entering lines in the initial, and departing lines in the final, coupling scheme.

(iii) Summation over the new quantum numbers which appear in the recoupling - here g and h - is implied in the graph.

The relation between the algebraic expression (1.4) and the corresponding graph fig.1.5 is one-to-one. This means that one can draw a graph and read off it the algebraic expression, or draw a graph for a given algebraic expression.

The basic recoupling transformation contains as special cases the recoupling of three angular momenta. They are represented by the basic graph of fig.1.5, where one of the angular momenta equals zero. There are four cases :

(i) a=0, fig.1.6

$$\left[\psi^{[b]}{}_{[\psi^{[c]}\psi^{[d]}]}{}^{[f]}\right]^{[i]} = \sum_h \begin{bmatrix} 0 & b & b \\ c & d & f \\ c & h & i \end{bmatrix} \left[\psi^{[c]}{}_{[\psi^{[b]}\psi^{[d]}]}{}^{[h]}\right]^{[i]} \quad ; \qquad (1.7)$$

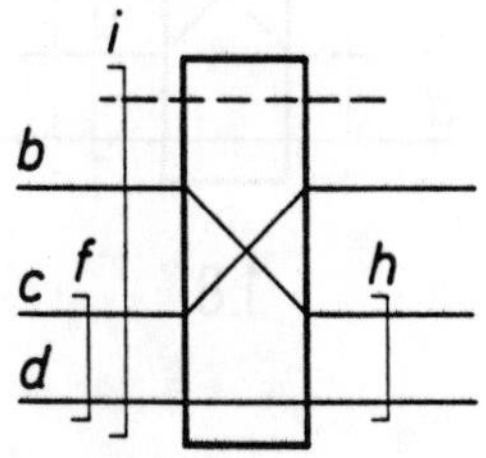

1.6

(ii) b=0, fig.1.7

$$\left[\psi^{[a]}{}_{[\psi^{[c]}\psi^{[d]}]}{}^{[f]}\right]^{[i]} = \sum_h \begin{bmatrix} a & 0 & a \\ c & d & f \\ h & d & i \end{bmatrix} \left[{}_{[\psi^{[a]}\psi^{[c]}]}{}^{[h]}\psi^{[d]}\right]^{[i]} \quad ; \qquad (1.8)$$

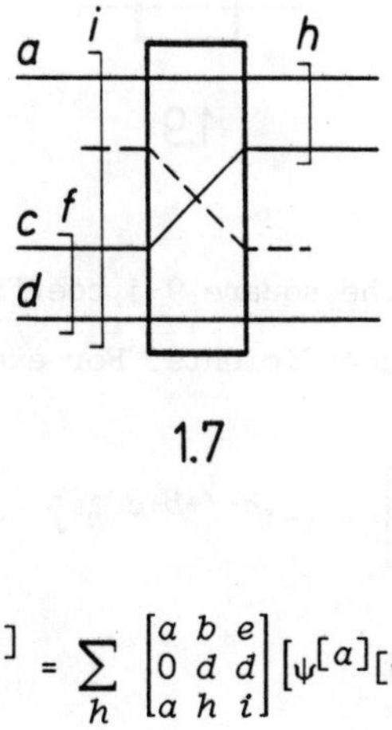

1.7

(iii) c=0, fig.1.8

$$\left[{}_{[\psi^{[a]}\psi^{[b]}\psi]}{}^{[e]}\psi^{[d]}\right]^{[i]} = \sum_h \begin{bmatrix} a & b & e \\ 0 & d & d \\ a & h & i \end{bmatrix} \left[\psi^{[a]}{}_{[\psi^{[b]}\psi^{[d]}]}{}^{[h]}\right]^{[i]} \quad ; \qquad (1.9)$$

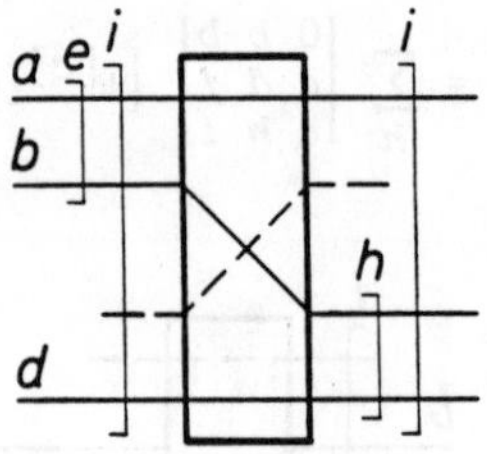

1.8

(iv) d=0, fig.1.9

$$\left[[{}_\psi[a]_\psi[b]][e]_\psi[c]\right]^{[i]} = \sum_h \begin{bmatrix} a & b & e \\ c & 0 & c \\ h & b & i \end{bmatrix} \left[[{}_\psi[a]_\psi[c]][h]_\psi[b]\right]^{[i]}. \qquad (1.10)$$

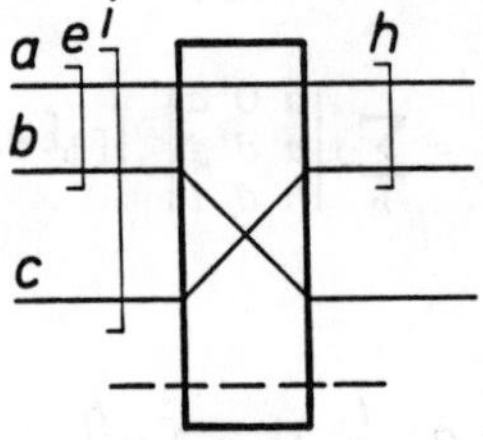

1.9

In these expressions the square 9-j coefficients with one zero are related to the Wigner 6-j coefficients. For example :

$$\begin{bmatrix} 0 & b & b \\ c & d & f \\ c & h & i \end{bmatrix} = (-)^{h+f+b+c}\,\hat{h}\,\hat{f}\,\begin{Bmatrix} i & h & c \\ d & f & b \end{Bmatrix}, \qquad (1.11)$$

or

$$\begin{bmatrix} a & 0 & a \\ c & d & f \\ h & d & i \end{bmatrix} = (-)^{i+c+a+d}\,\hat{h}\,\hat{f}\,\begin{Bmatrix} i & h & d \\ c & f & a \end{Bmatrix}. \qquad (1.12)$$

It is preferable not to use relations such as (1.11) and (1.12) when writing formal expressions because they carry phases and normalization constants which are confusing. In order to avoid such trivial complications in the expressions we shall keep the square 9-j notation and its graphical representation throughout. Thus recoupling of three tensors is always the basic recoupling box of fig.1.5 with a zero line, corresponding to a 9-j coefficient with a zero. At any rate, if one prefers to introduce 6-j coefficients, the complete list of conversion expressions is given in table 1, Chapter 10.

The change in the coupling order of two angular momenta

$$[{}_{\psi}[a]_{\psi}[b]][c] = (-)^{a+b-c}\ [{}_{\psi}[b]_{\psi}[a]][c] \tag{1.13}$$

is represented by the graph of fig.1.10.

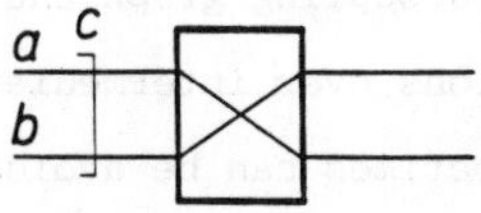

1.10

It is, of course, a special case of the basic recoupling box having two zero angular momenta

$$\begin{bmatrix} 0 & a & a \\ b & 0 & b \\ b & a & c \end{bmatrix} = (-)^{a+b-c}, \tag{1.14}$$

shown in fig.1.11.

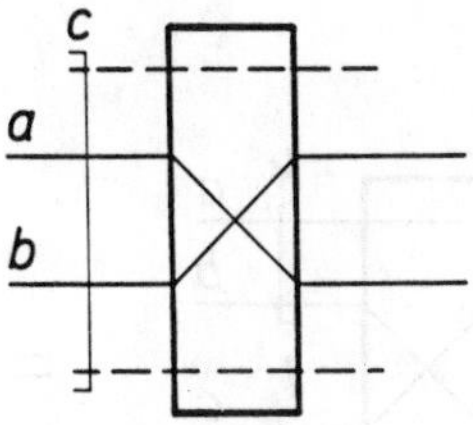

1.11

More generally, when the square 9-j coefficients which are read off from a graph contain more than one zero they are simple products of square roots and phases. All these cases are collected in table 2 (two zeroes) and table 3 (three zeroes) of Chapter 10.

Here we only note the values for the frequently occuring coefficients

$$\begin{bmatrix} a & a & 0 \\ b & b & 0 \\ c & c & 0 \end{bmatrix} = \begin{bmatrix} a & b & c \\ a & b & c \\ 0 & 0 & 0 \end{bmatrix} = \frac{\hat{c}}{\hat{a}\,\hat{b}} \,. \tag{1.15}$$

Identities between recouplings

After having drawn a recoupling graph one may find that the resulting expression contains summations over intermediate angular momenta. Some or all of these summations sometimes can be avoided in view of the existence of identities between recoupling schemes. This reduction is hard to carry out algebraically. However, it frequently can be done simply by re-drawing the graph.

In general, drawing several different recoupling graphs which bring a set of lines in a given coupling scheme to the same final coupling scheme, is a method for obtaining theorems on recoupling identities.

As a first example the orthogonality relation of the square 9-j coefficients is demonstrated by comparing fig.1.12a and fig.1.12b.

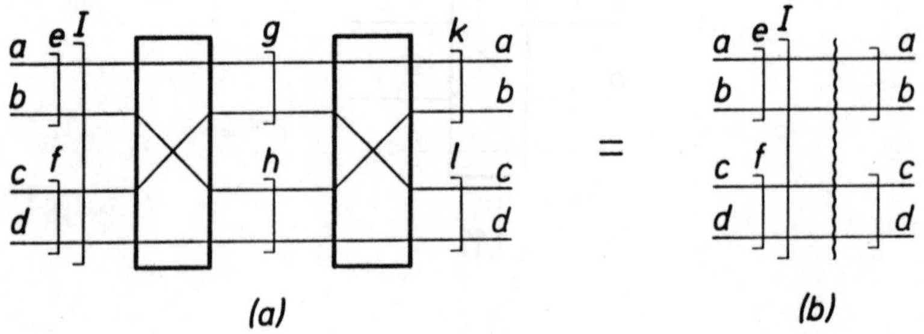

1.12

We read off

$$\sum_{g,h} \begin{bmatrix} a & b & e \\ c & d & f \\ g & h & I \end{bmatrix} \begin{bmatrix} a & c & g \\ b & d & h \\ k & \ell & I \end{bmatrix} = \delta_{ek}\delta_{f\ell}. \tag{1.16}$$

As a next example, the identity between the recouplings of fig.1.13(a) and 1.13(b) yields the identity

$$\begin{aligned} &\sum_g (-)^{a+b-e} \begin{bmatrix} b & a & e \\ c & 0 & c \\ g & a & I \end{bmatrix} (-)^{b+c-g} \begin{bmatrix} c & b & g \\ a & 0 & a \\ k & b & I \end{bmatrix} (-)^{c+a-k} \\ &= \sum_g (-)^{2a+g-e-k} \begin{bmatrix} b & a & e \\ c & 0 & c \\ g & a & I \end{bmatrix} \begin{bmatrix} c & b & g \\ a & 0 & a \\ k & b & I \end{bmatrix} \\ &= \begin{bmatrix} a & b & e \\ c & 0 & c \\ k & b & I \end{bmatrix} . \end{aligned} \tag{1.17}$$

In simplifying the phase we have utilized the triangularity condition which imposes that b+c-g, etc, are integer.

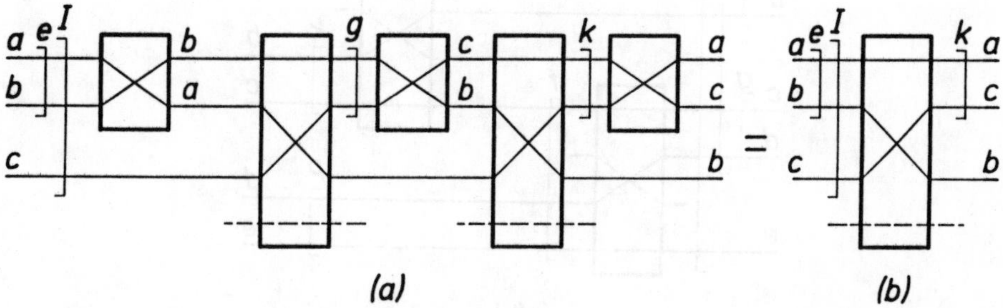

1.13

The somewhat more involved graphs of figs 1.14(a) and 1.14(b) wich couple five tensors to total angular momentum zero lead to the Biedenharn identity

$$\sum_{h} \begin{bmatrix} a & b & f \\ 0 & g & g \\ a & h & e \end{bmatrix} \begin{bmatrix} b & 0 & b \\ c & d & g \\ k & d & h \end{bmatrix} \begin{bmatrix} k & d & h \\ 0 & e & e \\ k & p & a \end{bmatrix}$$

$$= \begin{bmatrix} c & d & g \\ 0 & e & e \\ c & p & f \end{bmatrix} \begin{bmatrix} a & b & f \\ 0 & c & c \\ a & k & p \end{bmatrix}. \qquad (1.18)$$

In drawing the graph we have used the fact that the coupling to total angular momentum zero entails a symmetry which will be discussed in Chapter 4, and is given in eq. (4.2).

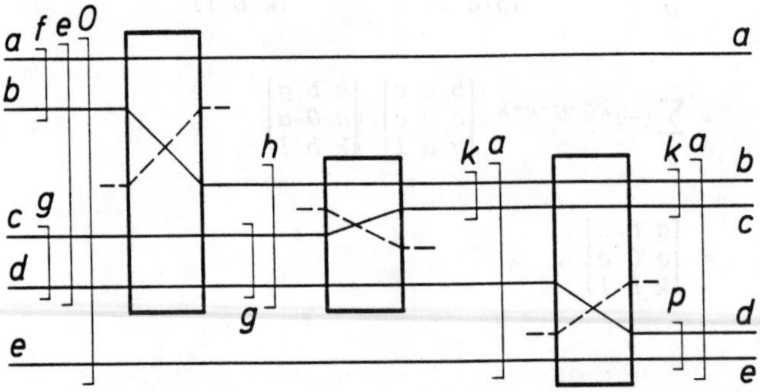

(a)

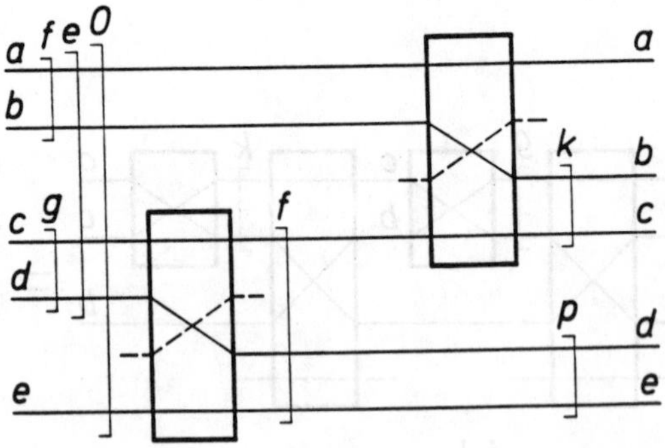

(b)

1.14

We will not need to remember any of these identities. They are meant to illustrate that in complicated recoupling transformations the number of dummy summations may be decreased by choosing judiciously the order of recouplings when drawing a graph.

In conclusion, we have seen that the use of the graph has many advantages. It helps finding the most direct and economical intermediate recouplings. It yields immediately the selection rules for intermediate angular momenta. It permits to write the algebraic expressions in a way which clarifies the physical meaning of the various factors. And last but not least, the presentation of a graph together with the final algebraic expression permits both to avoid the writing of intermediate expansions while still enabling one to verify the final results by directly reading the graph.

The Invariant Graph Method is immediately applicable for recoupling problems in SU(n) ; only one needs more quantum numbers to fully specify an irreducible representation (the quantum numbers a, b, c, etc. in e.g. eq.(1.4) become n-tuplets). The unchanged graph symbols now have values specific to SU(n). Otherwise, all rules for drawing and reading graphs remain unchanged.

Chapter 2

INVARIANTS

All angular momentum calculations are extremely simple if from the start one takes into account the fact that all physical quantities are invariants, i.e. they are independent of the coordinate system one chooses for their description. This is the case for norms, energies, cross sections, etc.. We choose to write also the wave functions themselves as invariants by introducing state amplitudes. To that end in this Chapter we first define the scalar product of two tensors. Then we use it to define the invariant form of wave functions by introducing the state amplitudes which specify the orientation of the physical objects. Then, using graphs we perform the computation for some examples of matrix elements, viz. norms and expectation values of scalar operators. By means of these examples we can complete the description of the basic steps of the Invariant Graph method.

Invariant tensor products

A quantity which is independent of the coordinate system is a scalar. In other words, invariants are zero angular momentum quantities. That means that the tensors they may contain are coupled to zero overall angular momentum. For example, in view of the expression for the vector coupling coefficients

$$(II\text{-}MM|00) = \frac{(-)^{I+M}}{\sqrt{2I+1}}, \tag{2.1}$$

an invariant containing two tensors is of the form

$$S^{[0]} = \left[\tilde{A}^{[I]}B^{[I]}\right]^{[0]} = \sum_M (II\text{-}MM|00)\ \tilde{A}^{[I]}_{-M}B^{[I]}_{M}$$

$$= \frac{1}{\hat{I}}\sum_M (-)^{I+M}\ \tilde{A}^{[I]}_{-M}B^{[I]}_{M}. \tag{2.2}$$

The tilda symbol means that in the cases of matrix or spinor tensors the transpose operation has been applied. We have here employed the shorthand notation $\hat{I}$, already introduced previously, eq.(1.6), which always will be used in this book

$$\hat{I} = \sqrt{2I+1}. \tag{2.3}$$

The expression (2.2) can be re-written in terms of the hermitian conjugate tensor of $A^{[I]}$, as defined in the Introduction, eq.(I.1),

$$A^{[I]\dagger}_{M} = (-)^{I+M}\ \tilde{A}^{[I]}_{-M}, \tag{2.4}$$

which yields

$$S^{[0]} = \frac{1}{\hat{I}}\sum_M A^{[I]\dagger}_{M}B^{[I]}_{M}. \tag{2.5}$$

This is the familiar definition of an invariant.

This way we see that an invariant in quantum mechanics can appear in the two forms : as the form (2.2) which is the coupling of two tensors to zero angular momentum, and as a bilinear form involving the product of a tensor and its hermitian conjugate, eq.(2.5). As a consequence of our phase convention eq. (2.4) these two forms are identical, with no relative phase.

We now illustrate these observations in some more detail by discussing in turn : i) the notion of invariants using the well-known example of wave function norms ; ii) the hermitian conjugation phase which ensues ; iii) the invariant form of wave functions describing an object of given polarization. This introduces the definition of the amplitudes $W_M^{[I]}$ which describe the preparation of the system.

Invariant form of states

A wave function commonly is written in the form

$$\psi_I = \sum_M W_M \psi_M^{[I]}. \tag{2.6}$$

The amplitudes W_M describe the preparation of the state. Each component $\psi_M^{[I]}$ is normalized according to

$$\int \psi_M^{[I]+} \psi_M^{[I]} = 1, \tag{2.7}$$

while the amplitudes are normalized according to

$$\sum_M |W_M|^2 = 1. \tag{2.8}$$

The properties of the state ψ_I being independent of the coordinate system, the expression (2.6) must be constructed as an invariant. Namely it must be given the form of the coupling to angular momentum zero of two tensors, viz. $W^{[I]}$ and $\psi^{[I]}$

$$\psi_I = \hat{I} \sum_M (II\text{-}MM|00)\ W_{-M}^{[I]} \psi_M^{[I]}, \tag{2.9}$$

or in compact notation this invariant form writes

$$\psi_I = \hat{I}\,[W^{[I]}\psi^{[I]}]^{[0]}. \tag{2.10}$$

We introduce now two quantities which shall be basic to all further developments, the invariant norm of the state components and the invariant norm or overlap of the state amplitudes.

Consider the norm of a single component

$$\langle\psi_M^{[I]}|\psi_M^{[I]}\rangle = \int \psi_M^{[I]+}\psi_M^{[I]} = 1. \tag{2.11}$$

Substituting the phase definition eq.(2.4) for the hermitian conjugation, introducing the value of the vector coupling coefficients, eq.(2.1), and summing over M, we get

$$\int \sum_M \psi_M^{[I]+}\psi_M^{[I]} = 2I+1 = \hat{I}\int \sum_M (II\text{-}MM|00)\,\tilde{\psi}_{-M}^{[I]}\psi_M^{[I]}$$

$$= \hat{I}\int [\tilde{\psi}^{[I]}\psi^{[I]}]^{[0]}. \tag{2.12}$$

Here and later the integral signifies integration over the coordinates and summation over spinor indices.

Because of its importance we introduce the symbol

$$\left[\psi^{[I]}\middle|\psi^{[I]}\right] = \int [\tilde{\psi}^{[I]}\psi^{[I]}]^{[0]} = \hat{I} \tag{2.13}$$

and we call it "the invariant norm". Hence, from eq.(2.12), we have the relation between the norm (2.11) and the invariant norm

$$\langle\psi_M^{[I]}|\psi_M^{[I]}\rangle = \frac{1}{\hat{I}}\left[\psi^{[I]}\middle|\psi^{[I]}\right]. \tag{2.14}$$

Note that the bra-ket notation [|] in the left hand side of eq.(2.13) implies the hermitian conjugation for the bra. Therefore in the notation of the invariant matrix element the tilda on the bra is dropped since it is redundant. The graph symbol for the invariant norm (2.13) is shown in fig.2.1.

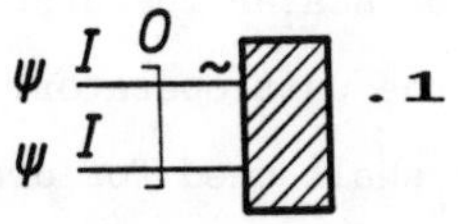

2.1

In a same way we introduce the invariant overlap for the amplitudes $W^{[I]}_{M}$. Using their normalization (2.8) we obtain

$$[W^{[I]\dagger}W^{[I]}]^{[0]} = \frac{1}{\hat{I}}. \qquad (2.15)$$

Note that the amplitudes $W^{[I]}_{M}$ obey our general phase convention. Again we introduce a bra-ket symbol for the invariant amplitude norm

$$[W^{[I]}|W^{[I]}] = [W^{[I]\dagger}W^{[I]}]^{[0]} = \frac{1}{\hat{I}}. \qquad (2.16)$$

The graph symbol of the invariant amplitude norm (2.16) is given in fig.2.2.

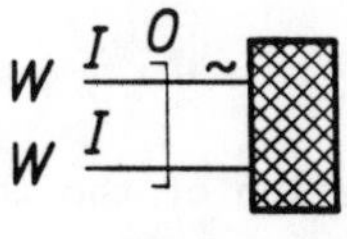

2.2

The amplitude overlap end-box fig.2.2, which has the value $\frac{1}{\hat{I}}$, is distinguished from the wave function norm end-box fig. 2.1, which has the value $\hat{I}$, by cross-hatching.

Computing matrix elements with graphs

Once the invariant form of the wave functions has been introduced, a graph can then be drawn for the calculation of matrix elements. A few simple examples involving only two tensors are given. These examples will include the matrix elements of scalar operators, such as the energy, as well as the introduction of the unit operator. These examples contain practically all the rules we shall need for drawing and evaluating graphs.

Norm of the invariant state

We evaluate the norm of the invariant state

$$\psi_I = \hat{I}\ [W^{[I]}\psi^{[I]}]^{[0]}. \tag{2.17}$$

It is

$$\begin{aligned}\langle\psi_I|\psi_I\rangle &= \hat{I}^2 \int [W^{[I]}\psi^{[I]}]^{[0]+}[W^{[I]}\psi^{[I]}]^{[0]}\\ &= \hat{I}^2 \int [\tilde{W}^{[I]}\ \tilde{\psi}^{[I]}]^{[0]}[W^{[I]}\psi^{[I]}]^{[0]}\\ &= \hat{I}^2\ \big[[W^{[I]}\psi^{[I]}]^{[0]}\big|[W^{[I]}\psi^{[I]}]^{[0]}\big].\end{aligned} \tag{2.18}$$

Here we have generalized the notation of the invariant matrix element, i.e. [|], to composite tensors. We again delete for economy the tildas in the bra of the final form since it is redundant with the bra notation. This expression will be evaluated by drawing a graph, fig.2.3.

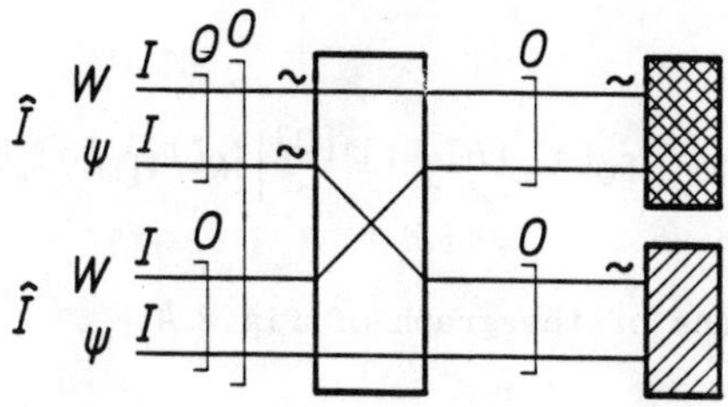

2.3

The drawing of the graph 2.3 proceeds in the following way. The expression (2.18) is represented by the left hand side of the graph. Reading in (2.18) the tensors from left to right, we represent them on the graph by the lines from top to bottom. The couplings are indicated. The calculation is graphically performed by recoupling the amplitudes and the wave functions into their respective invariant matrix elements. These matrix elements are represented by the end-boxes of the graph. Thus for the amplitude invariant overlap, eq.(2.16), we have the end-box symbol of fig.2.2, and for the wave function invariant norm, eq. (2.13), the symbol of fig.2.1. The tildas on the bra lines in the graph 2.3 are shown to indicate that they must enter the end-boxes at the top.

We may now compute the value of the norm (2.18). The result is simply the product of the values of all the symbols and factors of the graph 2.3. We read off :

$$\langle\psi_I|\psi_I\rangle = \hat{I}^2 \begin{bmatrix} I & I & 0 \\ I & I & 0 \\ 0 & 0 & 0 \end{bmatrix} \frac{1}{\hat{I}}\,\hat{I} = 1.$$

The value of the recoupling coefficient is a special case of eq.(1.15).

Norm of a 2-body wave function

We compute the norm of a two-body wave function in its invariant form constructed with normalized one-body wave functions,

$$\Psi_I = \hat{I}\,\left[W^{[I]}[\psi^{[a]}\psi^{[b]}]^{[I]}\right]^{[0]}. \tag{2.19}$$

The norm writes

$$\langle\Psi_I|\Psi_I\rangle = \hat{I}^2\left[\left[W^{[I]}[\psi^{[a]}\psi^{[b]}]^{[I]}\right]^{[0]}\middle|\left[W^{[I]}[\psi^{[a]}\psi^{[b]}]^{[I]}\right]^{[0]}\right] \qquad (2.20)$$

It is evaluated by means of the graph of fig.2.4.

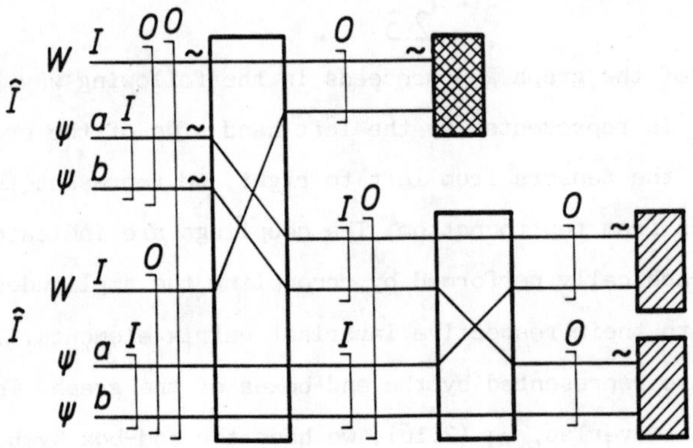

2.4

We note that the end-boxes, being invariants, impose the value zero on all new angular momenta appearing as the result of recoupling. We read off the graph 2.4 the product of all factors and symbols

$$\langle\Psi_I|\Psi_I\rangle = \hat{I}^2 \begin{bmatrix} I & I & 0 \\ I & I & 0 \\ 0 & 0 & 0 \end{bmatrix} \frac{1}{\hat{I}} \begin{bmatrix} a & b & I \\ a & b & I \\ 0 & 0 & 0 \end{bmatrix} \hat{a}\,\hat{b}$$

$$= \hat{I}^2 \frac{1}{\hat{I}^2}\frac{1}{\hat{I}}\frac{\hat{I}}{\hat{a}\,\hat{b}}\hat{a}\,\hat{b} = 1, \qquad (2.21)$$

which, as to be expected, indeed yields unity.

Here we have utilized the rule, which will prove to be one of the major advantages of the Invariant Graph method : namely, looking at the right-hand side of the graph we have recognized that the invariant end-boxes impose the value zero on all the new intermediate angular momenta arising at the exit of a recoupling box from the summation, eq.(1.4), of the recoupling transformation. Quite generally, selection rules for intermediate angular momenta will always be apparent on the graph and thus can be introduced directly in the algebraic expression.

Matrix element of a scalar operator

The matrix element of a scalar operator, for example the energy H, for a state ψ_I, eq.(2.17), writes

$$E = \hat{I}^2 \left[[W^{[I]}\psi^{[I]}]^{[0]}\right| H \left|[W^{[I]}\psi^{[I]}]^{[0]}\right]. \tag{2.22}$$

It has the same formal structure as the invariant norm, eq.(2.18). It is described by the simple graph 2.5 which yields,

$$E = \hat{I}^2 \begin{bmatrix} I & I & 0 \\ I & I & 0 \\ 0 & 0 & 0 \end{bmatrix} \frac{1}{\hat{I}} \left[\psi^{[I]}\right| H \left|\psi^{[I]}\right] = \frac{1}{\hat{I}} \left[\psi^{[I]}\right| H \left|\psi^{[I]}\right]. \tag{2.23}$$

Here we have introduced the notation for the invariant matrix element of an operator, which will be defined in full generality in Chapter 4.

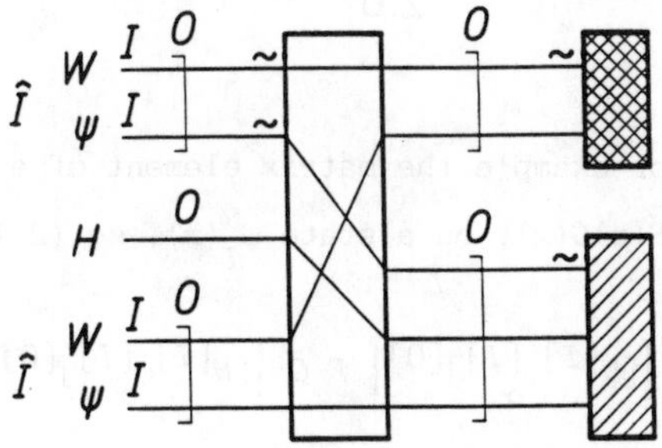

2.5

Insertion of the unit operator

In the calculation of matrix elements it is often necessary or convenient to introduce the unit operator.

For example, $\delta^3(x_1 - x_2)$ expressed in position space in terms of a complete set of orthonormal states will be written as a sum of invariant products

$$\delta^3(x_1-x_2) = \sum_{\alpha,j} \hat{j}\ [\tilde{\psi}^{[j]}_{\alpha}(x_1)\ \psi^{[j]}_{\alpha}(x_2)]^{[0]}, \tag{2.24}$$

where α denotes all the quantum numbers besides j needed to specify the state.

The graphical representation of the unit operator is given in fig.2.6. It is a box with outgoing lines. As always summation over new quantum numbers, here α and j, is implied. The box from which the lines originate yields in the final algebraic expressions the value $\hat{j}$, from eq.(2.24). Also, the tilda line, associated with a bra, emerges at the top.

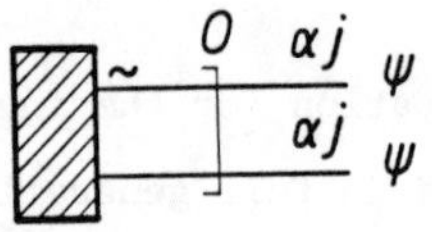

2.6

Let us evaluate for example the matrix element of a product of two scalar operators, $\Omega = F(x)G(x)$, on a state $\psi_I(x)$, eq.(2.10),

$$M = \hat{I}^2 \left[[W^{[I]}\psi^{[I]}_{\alpha}]^{[0]}\right|\ F\ G\ \left|[W^{[I]}\psi^{[I]}_{\alpha}]^{[0]}\right]. \tag{2.25}$$

In order to evaluate this expression we draw the graph of fig.2.7 involving the unit operator. Note that in drawing the graph we have immediately replaced the summation index j in the expression of the unit operator, eq. (2.24), by the index I imposed by the selection rules emanating from the zero angular momentum end-boxes. Likewise all intermediate new quantum numbers again are specified by these selection rules.

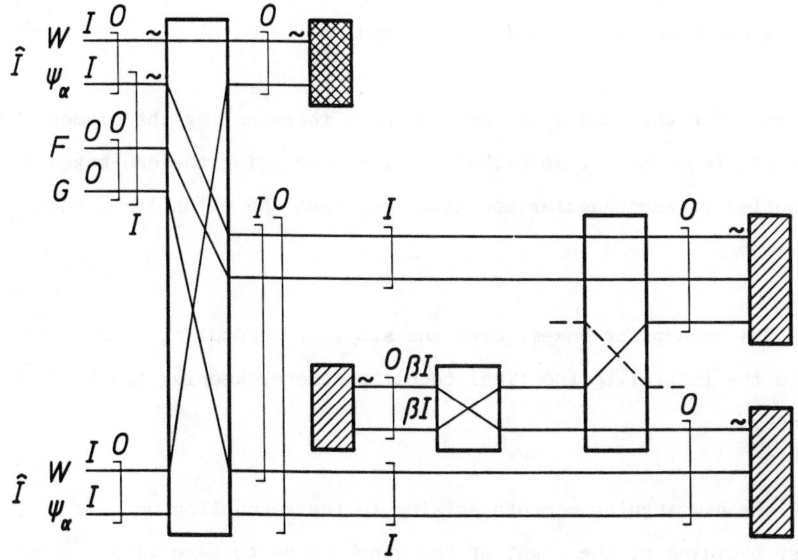

2.7

We read off from the graph the algebraic result

$$M = \hat{I}^2 \begin{bmatrix} I & I & 0 \\ I & I & 0 \\ 0 & 0 & 0 \end{bmatrix} \frac{1}{\hat{I}} \sum_{\beta} \hat{I}(-)^{2I} \begin{bmatrix} I & 0 & I \\ I & I & 0 \\ 0 & I & I \end{bmatrix} \left[\psi_{\alpha}^{[I]}\right| F \left|\psi_{\beta}^{[I]}\right] \left[\psi_{\beta}^{[I]}\right| G \left|\psi_{\alpha}^{[I]}\right]$$

$$= \frac{1}{\hat{I}^2} \sum_{\beta} \left[\psi_{\alpha}^{[I]}\right| F \left|\psi_{\beta}^{[I]}\right] \left[\psi_{\beta}^{[I]}\right| G \left|\psi_{\alpha}^{[I]}\right]. \tag{2.26}$$

The basic steps of the Invariant Graph method

Although trivial, these examples contain the basic steps of the Invariant Graph method, namely :

(1) Draw on the left-hand side from top to bottom the lines associated with the quantities in the initial invariant matrix element read from left to

right. Initial couplings are indicated by brackets labelled with the corresponding angular momenta. All tensor lines associated to the bra are denoted by a tilda. Write as shown on the left-hand side of this graph all the factors contained in the initial expression.

(2) Draw on the right hand side the end-boxes representing the elementary invariant matrix elements. Note that the lines entering the end-boxes are always coupled to zero angular momentum, and that the tilda lines enter at the top.

(3) Using 9-j recoupling boxes, draw the simplest recoupling graph which leads from the initial to the final coupling scheme, keeping track of the tilda lines.

(4) Label the new angular momenta arising in the recoupling or unit operators, begining at the right of the graph so as to take into account the zero total angular momentum selection rule imposed by each end-box on its entering lines .

(5) Read off the algebraic result from the graph by forming the product of the values of the elements of the graph. If the selection rules arising from the end-boxes leave some new quantum numbers open, sum over their values.

We now have made acquaintance with the basic elements of the Invariant Graph method, including practically all its rules.

It is seen that the simplicity of the method rests on the direct one-to-one correspondence between the graph and the final algebraic expression. The result is simply the product of the values of all the elements of the graph, without any further factors or phases. This direct correspondence is possible because we work only with wave functions in their invariant form, and also because we have adopted a phase convention for hermitian conjugation such that the phase of the ket to bra transformation needed in the definition of invariants generates the vector coupling coefficients to angular momentum zero.

To complete the presentation of the graphical method we only need to extend the definition of the invariant to non-scalar operators and their matrix elements. This is carried out in Chapter 4. However it is useful to first dwell in some more detail on the material developed so far.

Chapter 3

PHASES, WAVE FUNCTIONS, AND OPERATORS

This Chapter covers several disconnected subjects which complement Chapters 1 and 2. We begin by giving some mathematical background to the phase convention adopted in this book, in particular the definitions of standard and contrastandard tensors and their relation. This relation is used to write all quantities exclusively as contrastandard tensors. This is the key to the systematics of the Invariant Graph method. Next we illustrate the concept of invariant wave functions and operators introduced in Chapter 2 and we give some examples of state and operator amplitudes. To complete the Chapter we write the transformation from cartesian to spherical tensors and we give the scalar product of vectors as an example.

Contrastandard and standard tensors

All tensors of quantum physics, i.e. wave functions, amplitudes, operators, etc.., denoted for example by $\psi_M^{[I]}$, $A_M^{[I]}$, $\Omega_M^{[I]}$, $a_M^{[I]}$ etc... fulfill

$$\vec{I}^2\psi_M^{[I]} = I(I+1)\psi_M^{[I]}, \tag{3.1}$$

$$I_z\,\psi_M^{[I]} = M\,\psi_M^{[I]}, \tag{3.2}$$

where $\vec{I}$, I_z are the angular momentum operators. Furthermore the square bracket notation in the superscripts, i.e. $[I]$, denotes that these tensors are defined as obeying the contrastandard phase convention. This means that rotating the objects through the Euler angles α, β, γ the tensors transform as

$$A'^{[I]}_{M} = \sum_{M'=-I}^{I} A^{[I]}_{M'}\, \mathfrak{D}^{I}_{M'M}(\alpha,\beta,\gamma). \tag{3.3}$$

Here the $\mathfrak{D}^{I}_{M'M}$ are the Wigner rotation matrices.

The hermitian conjugation of a contrastandard tensor does three operations : (i) it performs time reversal, (ii) it performs the transpose operation on any spinor or matrix, (iii) by convention it introduces a phase. The hermitian conjugate of a contrastandard tensor is denoted with round parenthesis in the superscript, i.e. (I),

$$\Psi^{[I]\dagger}_{M} = \tilde{\Psi}^{(I)}_{M}\ , \tag{3.4}$$

to indicate that it is a standard tensor, i.e. which transforms under rotation as

$$A'^{(I)}_{M} = \sum_{M'=-I}^{I} A^{(I)}_{M'}\, \mathfrak{D}^{I*}_{M'M}(\alpha,\beta,\gamma). \tag{3.5}$$

The tilda in eq.(3.4) means that all matrices or spinors contained in the tensor undergo the transpose operation.

The relative phase between standard and contrastandard tensors is fixed by choosing

$$A^{(I)}_{M} = (-)^{I+M}\, A^{[I]}_{-M}\ . \tag{3.6}$$

The hermitian conjugation of a contrastandard tensor, eq.(3.4), can now be re-expressed in terms of a contrastandard tensor

$$\psi_M^{[I]\dagger} = (-)^{I+M} \tilde{\psi}_{-M}^{[I]} \tag{3.7}$$

This relation is the basic phase convention of the book, eq.(I.1). It will allow us to work exclusively with contrastandard tensors. Whenever standard tensors arise (for example bras) they always will be written in terms of contrastandard quantities using eqs.(3.4), (3.6) and (3.7). Working systematically with contrastandard tensors simplifies drastically the handling of hermitian conjugate quantities (bra) in matrix elements or, as we shall see later, the handling of hole or anti-particle operators in Fock space.

The tilda in (3.7) indicating the transpose operation can be omitted if $\psi_M^{[I]}$ does not contain spinor or matrix quantities.

For commuting tensors, the hermitian conjugation property (3.7) transfers to coupled tensors without change of order in the coupling scheme. For example for two tensors, eq.(1.1), we have

$$\left[A^{[J]} B^{[K]}\right]_M^{[I]\dagger} = (-)^{I+M} [\tilde{A}^{[J]} \tilde{B}^{[K]}]_{-M}^{[I]} . \tag{3.8}$$

The extension to non-commuting tensors is done in Chapter 7.

Spin-orbit wave functions

The invariant form (2.10) of a spin-orbit state is

$$\psi_j = \hat{j} \left[W^{[j]} \psi_{s\ell}^{[j]}\right]^{[0]} , \tag{3.9}$$

where the amplitudes $W_m^{[j]}$ describe the state. The spin-orbit component wave-functions

$$\psi_{s\ell m}^{[j]} = [\chi^{[s]} Y^{[\ell]}]_m^{[j]} \tag{3.10}$$

fulfill the conjugation property (3.8),

$$\psi^{[j]\dagger}_{s\ell m} = (-)^{j+m}\tilde{\psi}^{[j]}_{s\ell -m} = (-)^{j+m}\,[\tilde{\chi}^{[s]}\,\tilde{Y}^{[\ell]}]^{[j]}_{-m}, \tag{3.11}$$

and they are normalized. This can be verified from the definition and conjugation properties of the spin and orbital functions, given in the Introduction, eqs.(I.2) and (I.4), and by computing the norm along the lines of eq.(2.20). The tilda on the spherical harmonic function could have been omitted.

The case with spin $s = 1$ has special significance. The vector functions

$$\vec{Y}^{[J]}_{\ell M} = [e^{[1]}Y^{[\ell]}]^{[J]}_{M}, \tag{3.12}$$

written with the spin 1 wave functions $e^{[1]}$, are called the "vector spherical harmonics".

In order to fulfill the conjugation property and normalization we define the spin 1 wave functions $e^{[1]}$, which at the same time provide the spherical representation of the unit vector, in terms of the Cartesian unit vectors as

$$\begin{aligned} e^{[1]}_{1} &= i\,\frac{1}{\sqrt{2}}\,(\vec{e}_x+i\vec{e}_y), \\ e^{[1]}_{1} &= -i\vec{e}_z, \\ e^{[1]}_{-1} &= -i\,\frac{1}{\sqrt{2}}\,(\vec{e}_x-i\vec{e}_y). \end{aligned} \tag{3.13}$$

The cartesian unit vectors $\vec{e}_x$, $\vec{e}_y$, $\vec{e}_z$, fulfill $e_i^2 = 1$, $\vec{e}_i\cdot\vec{e}_j = 0$. These definitions again ensure the basic phase convention (3.7)

$$e^{[1]\dagger}_{M} = (-)^{1+M}\,e^{[1]}_{-M}, \tag{3.14}$$

where we have omitted the tilda.

Finally the jj- to LS-coupling transformation for a two-particle state

$$\psi_I = \hat{I}\left[W^{[I]}\,[\psi_a^{[j_a]}\psi_b^{[j_b]}]^{[I]}\right]^{[0]}, \tag{3.15}$$

is given by the graph fig.3.1 which yields

$$\psi_I = \hat{I}\sum_{SL}\begin{bmatrix} s_a & \ell_a & j_a \\ s_b & \ell_b & j_b \\ S & L & I \end{bmatrix}\left[W^{[I]}\left[[\chi_a^{[s_a]}\chi_b^{[s_b]}]^{[S]}[\varphi_a^{[\ell_a]}\,\varphi_b^{[\ell_b]}]^{[L]}\right]^{[I]}\right]^{[0]}. \tag{3.16}$$

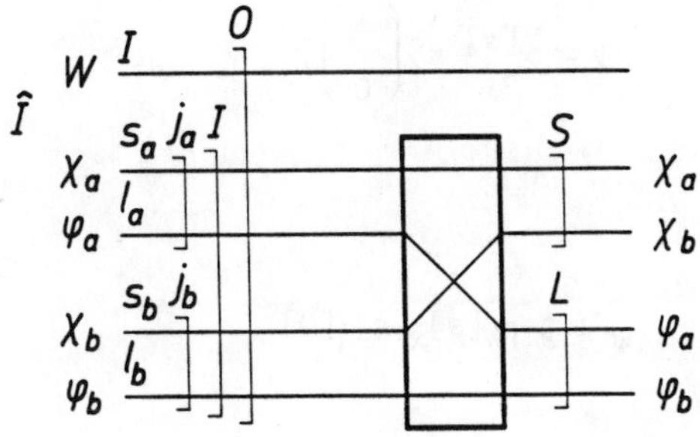

3.1

State amplitudes

We now show on a few examples how to construct invariant state vectors of given polarization with the amplitudes $W_M^{[I]}$ of Chapter 2. These amplitudes by a proper definition of their phases can be chosen to achieve invariant forms of definite reality character, i.e. a purely real or a purely imaginary form. The definition of invariant states of definite reality character can be helpful in performing transition matrix element calculations.

As a first example we consider the case of a real polarized orbital state with projection $m = 0$ along the Oz-axis. It is phased acording to :

$$\psi \simeq Y_{\ell 0} \simeq P_\ell(\cos\theta). \tag{3.17}$$

Therefore, in order for the invariant state

$$\psi_\ell = \hat{\ell}\, [W^{[\ell]} Y^{[\ell]}]^{[0]}, \tag{3.18}$$

to be real, we require that

$$W^{[\ell]}_m = (-i)^\ell\, \delta_{m0}. \tag{3.19}$$

Likewise for a polarized spin $s = ½$ state with $m = ½$ along the Oz-axis,

$$\psi \simeq \chi^{[s]}_{½} = \begin{pmatrix} 1 \\ 0 \end{pmatrix}, \tag{3.20}$$

we need for the form

$$\psi = \hat{s}\, [W^{[s]} \chi^{[s]}]^{[0]} \tag{3.21}$$

to be real that the amplitudes be given by

$$W^{[s]}_m = -\,\delta_{m,-½}. \tag{3.22}$$

Similarly, a state $s = ½$ polarized with $m = -½$ along the Oz-axis requires

$$W^{[s]}_m = \delta_{m,½}. \tag{3.23}$$

As the last example, in the case of spin 1, the polarization vector $\vec{P}$, defined in terms of the spherical unit vectors (3.13)

$$\vec{P} = \hat{1}\, [W^{[1]} e^{[1]}]^{[0]}, \tag{3.24}$$

is real if the components $W^{[1]}_M$ are defined as spherical tensors according to the expressions given below in eq. (3.31).

Invariant form of operators

Operators also are defined to be contrastandard tensors which obey the basic phase definition eq.(3.7),

$$\Omega_{\mu}^{[\lambda]+} = (-)^{\lambda+\mu}\,\tilde{\Omega}_{-\mu}^{[\lambda]}. \tag{3.25}$$

They always will be used in the invariant form

$$\Omega_{\lambda} = \hat{\lambda}\,[\omega^{[\lambda]}\Omega^{[\lambda]}]^{[0]}, \tag{3.26}$$

constructed with the amplitudes $\omega_{\mu}^{[\lambda]}$ which describe the physical process.

For example for a plane wave operator its multipole decomposition

$$e^{i\vec{k}.\vec{r}} = 4\pi \sum_{\ell} (-i)^{\ell} j_{\ell}(kr)\,\hat{\ell}\,[Y^{[\ell]}(\hat{k})\,Y^{[\ell]}(\hat{r})] \tag{3.27}$$

shows that the $\omega_{\mu}^{[\lambda]}$ are associated with measurement of the angular distributions and are given by the functions $Y_{\mu}^{[\lambda]}(\hat{k})$. The details are contained in Chapter 9.

Similarly, for a dipole operator with linear polarization along the Oz direction one has

$$\left|\omega_{m'}^{[1]}\right| = \delta_{m'0}. \tag{3.28}$$

We mention here that for the trivial but frequent case of a scalar operator there holds

$$\omega_{0}^{[0]} = 1. \tag{3.29}$$

Many examples of operator amplitudes will appear in the later chapters of this book.

Scalar product of vector operators

A special but important case of invariant operators is the scalar product of two vector operators, for example in spin-spin or spin-orbit interactions.

In order to write the scalar product of two cartesian vectors

$$\vec{A} \cdot \vec{B} = A_x B_x + A_y B_y + A_z B_z, \tag{3.30}$$

in the spherical representation, we introduce the components

$$V_1^{[1]} = i \frac{1}{\sqrt{2}} (V_x + iV_y),$$

$$V_0^{[1]} = -iV_z,$$

$$V_{-1}^{[1]} = -i \frac{1}{\sqrt{2}} (V_x - iV_y), \tag{3.31}$$

which obey our phase convention. Herewith the scalar product (3.30) becomes

$$\vec{A} \cdot \vec{B} = \hat{1} \, [A^{[1]} B^{[1]}]^{[0]}. \tag{3.32}$$

As always the invariant shows up as a zero angular momentum tensor.

Two important cases are the spin-spin interaction

$$\vec{\sigma}(1) \cdot \vec{\sigma}(2) = \hat{1} \, [\sigma^{[1]}(1) \sigma^{[1]}(2)]^{[0]}, \tag{3.33}$$

and the spin-orbit interaction

$$\vec{\sigma} \cdot \vec{L} = \hat{1} \, [\sigma^{[1]} L^{[1]}]^{[0]}, \tag{3.34}$$

where the vector tensors all are defined according to eq.(3.31) in terms of their cartesian components.

Chapter 4

MATRIX ELEMENTS

In order to evaluate matrix elements we introduce the invariant triple product of three tensors and discuss its properties. We then construct invariant matrix elements of tensor operators as integrals over invariant triple products. We thus obtain properly phased invariant matrix elements which then can be represented by end-boxes in the graphs without extraneous phases. We relate the Wigner-Eckart theorem to this definition of the invariant matrix element. This terminates the list of definitions necessary to use the graph method. Several applications to standard situations are given at the end of this Chapter.

The invariant triple product and its properties

The invariant constructed from three tensors, called the "invariant triple product", has the form

$$T^{[0]}= [A^{[I]}B^{[J]}C^{[K]}]^{[0]}. \tag{4.1}$$

In an invariant triple product the intermediate coupling does not need to be specified. This is demonstrated by the recoupling graph of fig.4.1 which reads

$$\left[\left[A^{[I]}B^{[J]}\right]^{[K]}C^{[K]}\right]^{[0]} = \begin{bmatrix} I & J & K \\ 0 & K & K \\ I & I & 0 \end{bmatrix} \left[A^{[I]}\left[B^{[J]}C^{[K]}\right]^{[I]}\right]^{[0]}$$

$$= \left[A^{[I]}\left[B^{[J]}C^{[K]}\right]^{[I]}\right]^{[0]}, \tag{4.2}$$

owing to (see table 2, Chapter 10)

$$\begin{bmatrix} I & J & K \\ 0 & K & K \\ I & I & 0 \end{bmatrix} = 1. \tag{4.3}$$

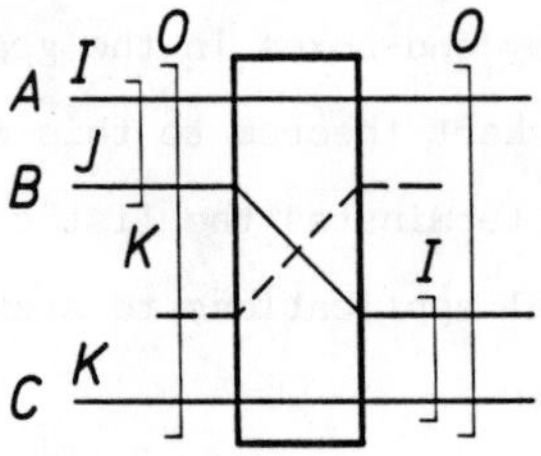

4.1

Consequently one does not have to indicate the intermediate angular momentum when representing the invariant triple product, fig.4.2.

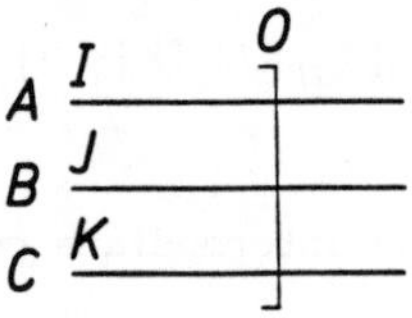

4.2

Applying the re-ordering relation (1.13) one finds the permutation properties valid for commuting tensors

$$[A^{[I]}B^{[J]}C^{[K]}]^{[0]} = (-)^{2K}\,[C^{[K]}A^{[I]}B^{[J]}]^{[0]}$$

$$= (-)^{J+K-I}[A^{[I]}C^{[K]}B^{[J]}]^{[0]} \text{ etc} \tag{4.4}$$

In that case, as a consequence of our phase convention, the hermitian conjugation of the invariant triple product yields

$$[A^{[I]}B^{[J]}C^{[K]}]^{[0]+} = [\tilde{A}^{[I]}\tilde{B}^{[J]}\tilde{C}^{[K]}]^{[0]}. \tag{4.5}$$

The invariant triple product of normalized tensors is normalized as shown by the norm graph fig.4.3 which yields, using eq.(1.15),

$$\mathcal{N} = \begin{bmatrix} K & K & 0 \\ K & K & 0 \\ 0 & 0 & 0 \end{bmatrix} \begin{bmatrix} I & J & K \\ I & J & K \\ 0 & 0 & 0 \end{bmatrix} \hat{I}\,\hat{J}\,\hat{K} = 1. \tag{4.6}$$

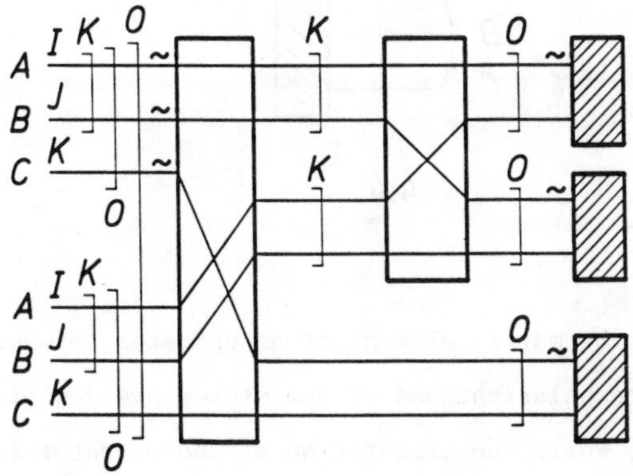

4.3

For completeness we give the explicit expression for the invariant triple product in terms of the Wigner 3-j coefficients

$$[A^{[I]}B^{[J]}C^{[K]}]^{[0]} = (-)^{I-J+K}$$

$$\times \sum_{M_I M_J M_K} \begin{pmatrix} I & J & K \\ M_I & M_J & M_K \end{pmatrix} A^{[I]}_{M_I} B^{[J]}_{M_J} C^{[K]}_{M_K}. \tag{4.7}$$

Invariant matrix elements

The invariant matrix elements involve the integration over the variables of an invariant triple product where the leftermost tensor is a bra, i.e. a Hermitian conjugate tensor. Thus we define the invariant matrix element as

$$\left[A^{[I]}\right| B^{[J]} \left|C^{[K]}\right] \equiv \int [\tilde{A}^{[I]} B^{[J]} C^{[K]}]^{[0]}. \tag{4.8}$$

It follows that the invariant matrix element has all the symmetries of the invariant triple product unless its three constituent quantities are non-commuting. The graphical representation of this invariant matrix element is given in fig. 4.4.

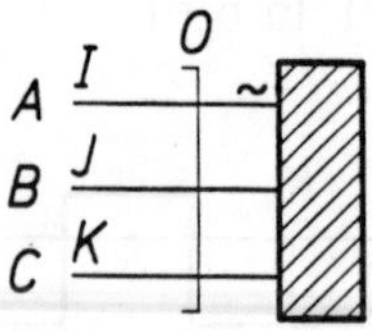

4.4

We now consider the matrix element of an operator between an initial and a final state. The polarizations of the states are described by amplitudes W_i and W_f, while the orientation of the operator is described by amplitudes ω. These amplitudes are used according to the definition of eq.(2.10) and eq.(3.26) for writing the states and the operator respectively as invariants. We thus have to evaluate the matrix element

$$\begin{aligned} T_{fi} &= \langle \psi_{If} | \Omega_\lambda | \psi_{Ji} \rangle \\ &= \hat{I}\,\hat{\lambda}\,\hat{J}\, \left[[W_f^{[I]} \psi_f^{[I]}]^{[0]} \middle| [\omega^{[\lambda]} \Omega^{[\lambda]}]^{[0]} \middle| [W_i^{[J]} \psi_i^{[J]}]^{[0]} \right] . \end{aligned} \tag{4.9}$$

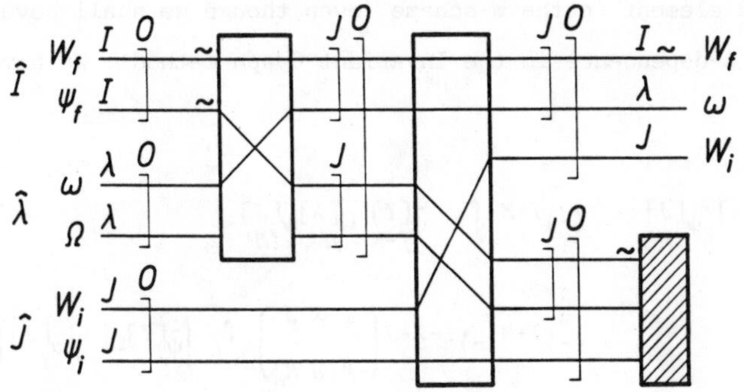

4.5

The corresponding graph, fig.4.5, yields

$$T_{fi} = \hat{I}\ \hat{\lambda}\ \hat{J} \begin{bmatrix} I & I & 0 \\ \lambda & \lambda & 0 \\ J & J & 0 \end{bmatrix} \begin{bmatrix} J & J & 0 \\ J & J & 0 \\ 0 & 0 & 0 \end{bmatrix} \left[\psi_f^{[I]}\middle|\Omega^{[\lambda]}\middle|\psi_i^{[J]}\right] [\tilde{W}_f^{[I]}\omega^{[\lambda]}W_i^{[J]}]^{[0]}, \tag{4.10}$$

which simplifies to

$$T_{fi} = [\tilde{W}_f^{[I]}\omega^{[\lambda]}W_i^{[J]}]^{[0]} \left[\psi_f^{[I]}\middle|\ \Omega^{[\lambda]}\ \middle|\psi_i^{[J]}\right]. \tag{4.11}$$

As it will always be the case, the result factorizes into an (or several) invariant matrix element describing the dynamics, and the amplitude factor, here $[\tilde{W}_f^{[I]}\omega^{[\lambda]}W_i^{[J]}]^{[0]}$, reflecting the particular experimental situation. This factor yields the values of angular distributions, polarizations, correlations, etc. In Chapter 9 we shall see how this factor is evaluated for some particular experiments.

The special case of a scalar operator, for example the energy, has $\lambda=0$ and $\omega^{[0]} = 1$ and its matrix element was calculated in Chapter 2, eq.(2.23).

The Wigner-Eckart theorem

We now give for completeness the relation between the invariant matrix element (4.8) and the usual Wigner reduced matrix element. To this end we write the matrix element in the m-scheme (even though we shall never need explicitly this m-dependence in the Invariant Graph method). We have with (4.7)

$$\int \psi^{[I]\dagger}_{fM}\, \Omega^{[\lambda]}_{\mu}\, \psi^{[J]}_{iM'} = (-)^{I+M} \int \tilde{\psi}^{[I]}_{f-M}\, \Omega^{[\lambda]}_{\mu} \psi^{[J]}_{iM'}$$

$$= (-)^{I+M}(-)^{I-\lambda+J}\begin{pmatrix} I & \lambda & J \\ -M & \mu & M' \end{pmatrix} \int \left[\tilde{\psi}^{[I]}_{f}\Omega^{[\lambda]}\psi^{[J]}_{i}\right]^{[0]}$$

$$= (-)^{I+M}(-)^{I-\lambda+J}\begin{pmatrix} I & \lambda & J \\ -M & \mu & M' \end{pmatrix} \left[\psi^{[I]}_{f}\right|\, \Omega^{[\lambda]}\, \left|\psi^{[J]}_{i}\right]. \qquad (4.12)$$

where we have used in the last line the notation (4.8) for denoting the invariant matrix element of the operator

$$\left[\psi^{[I]}_{f}\middle|\Omega^{[\lambda]}\middle|\psi^{[J]}_{i}\right] = \int \left[\tilde{\psi}^{[I]}_{f}\, \tilde{\Omega}^{[\lambda]}\psi^{[J]}_{i}\right]^{[0]} \qquad (4.13)$$

The basis of the relations in (4.12) is provided by the observation that the integration symbol on the left-hand side of (4.12) projects on the overall angular momentum zero component of its integrand.

The equation (4.12) is the Wigner-Eckart theorem, written however with the proper phase such that the coupling to overall zero angular momentum is performed with the phase conventions of Chapter 2. From this equation it is seen that the invariant matrix element (4.13) is related to the usual Wigner reduced matrix element by

$$\left[\psi^{[I]}_{f}\right|\, \Omega^{[\lambda]}\, \left|\psi^{[J]}_{i}\right] = (-)^{I+\lambda-J}\left\langle\psi^{[I]}_{f}\right\|\, \Omega^{[\lambda]}\, \left\|\psi^{[J]}_{i}\right\rangle. \qquad (4.14)$$

Evaluation of elementary invariant matrix elements.

The basic invariant matrix elements needed are in fact very few. They are given in Chapter 10 for the most common operators of quantum physics. They are computed by the usual method, i.e. one performs the calculation for specific values of the m-quantum numbers.The theorem (4.12) can then be used to evaluate specific invariant matrix elements. As an example we compute the invariant matrix element of the spin operator $\vec{\sigma}$ between spin ½ states.

The correctly phased spin matrix $\vec{\sigma}$ has according to eq.(3.31) the components

$$\sigma^{[1]}_{1} = i\,\frac{1}{\sqrt{2}}\,(\sigma_x + i\sigma_y),$$

$$\sigma^{[1]}_{0} = -i\sigma_z,$$

$$\sigma^{[1]}_{-1} = -i\,\frac{1}{\sqrt{2}}\,(\sigma_x - i\sigma_y),$$

$$(4.15)$$

with the usual representation for the σ_i

$$\sigma_x = \begin{pmatrix} 0 & 1 \\ 1 & 0 \end{pmatrix} \;;\; \sigma_y = \begin{pmatrix} 0 & -i \\ i & 0 \end{pmatrix} \;;\; \sigma_z = \begin{pmatrix} 1 & 0 \\ 0 & -1 \end{pmatrix}. \tag{4.16}$$

We have,

$$\chi^{[s]+}_{\frac{1}{2}} . \sigma^{[1]}_{0} . \chi^{[s]}_{\frac{1}{2}} = (-i)(1\;0)\begin{pmatrix} 1 & 0 \\ 0 & -1 \end{pmatrix}\begin{pmatrix} 1 \\ 0 \end{pmatrix} = -i.$$

On the other hand, using eq.(4.12), we also get

$$\chi^{[s]+}_{\frac{1}{2}} . \sigma^{[1]}_{0} . \chi^{[s]}_{\frac{1}{2}} = (-)^{s+\frac{1}{2}}(-)^{s-1+s}\begin{pmatrix} \frac{1}{2} & 1 & \frac{1}{2} \\ -\frac{1}{2} & 0 & \frac{1}{2} \end{pmatrix}[\chi^{[s]}|\sigma^{[1]}|\chi^{[s]}]. \tag{4.17}$$

Herewith we obtain

$$[\chi^{[s]}|\,\sigma^{[1]}\,|\chi^{[s]}] = i\sqrt{6}. \tag{4.18}$$

In a similar manner one can obtain the elementary invariant matrix element of the angular momentum operator

$$[\psi^{[I]}|L^{[1]}|\psi^{[I]}] = i\,\hat{I}\,\sqrt{I(I+1)}, \tag{4.19}$$

and of the spherical harmonic operator

$$[Y^{[\ell_1]}|Y^{[\ell_2]}|Y^{[\ell_3]}] = [\ell_1|\ell_2|\ell_3] = (-)^{(\ell_1+\ell_2+\ell_3)/2}\,\frac{\hat{\ell}_1\hat{\ell}_2\hat{\ell}_3}{\sqrt{4\pi}}\begin{pmatrix}\ell_1 & \ell_2 & \ell_3\\ 0 & 0 & 0\end{pmatrix}. \tag{4.20}$$

The list of the most frequently used (in fact essentially all) elementary invariant matrix elements is given in Chapter 10.

Examples

The spin-orbit interaction

We evaluate the matrix element of the spin-orbit operator

$$\vec{s}\cdot\vec{L} = \frac{1}{2}\,\vec{\sigma}\cdot\vec{L}, \tag{4.21}$$

for a state in j-j coupling,

$$\psi_j = \hat{j}\,[W^{[j]}\psi^{[j]}_{s\ell}]^{[0]}. \tag{4.22}$$

The invariant form of the operator is, eq.(3.34)

$$\vec{s}\cdot\vec{L} = \frac{1}{2}\,\hat{1}\,[\sigma^{[1]}L^{[1]}]^{[0]}. \tag{4.23}$$

Herewith the complete matrix element we have to compute is

$$M = \langle\psi_j|\frac{1}{2}\,\vec{\sigma}\cdot\vec{L}|\psi_j\rangle$$

$$= \hat{j}^2\,\frac{1}{2}\,\hat{1}\,\left[[W^{[j]}\psi^{[j]}_{s\ell}]^{[0]}\right|\,[\sigma^{[1]}L^{[1]}]^{[0]}\,\left|[W^{[j]}\psi^{[j]}_{sk}]^{[0]}\right]. \tag{4.24}$$

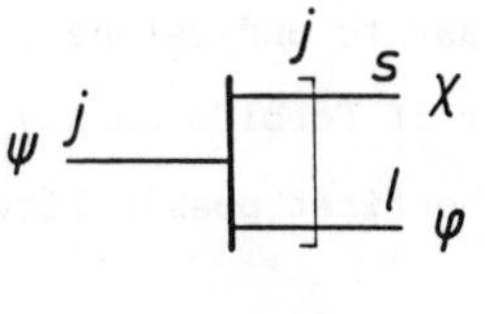

4.6

The graph associated with this expression for clarity will make use of the notation for coupled tensors shown in fig.4.6 which is self-explanatory. The complete graph for the matrix element (4.24) is shown in fig.4.7

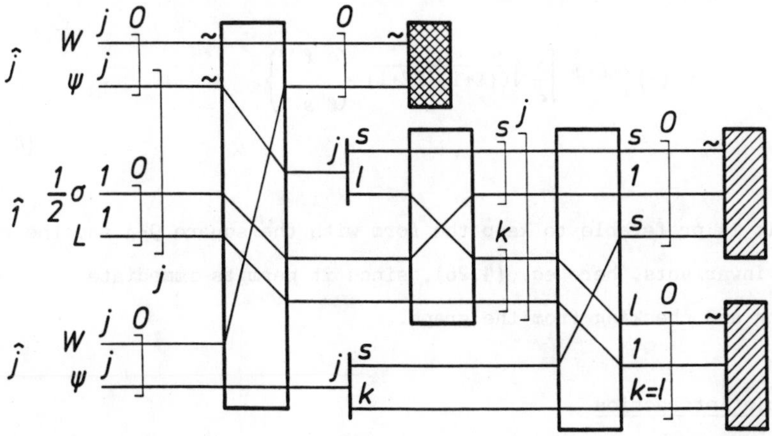

4.7

We read off :

$$M = \hat{j}^2\,\frac{1}{2}\,\hat{1}\begin{bmatrix} j & j & 0\\ j & j & 0\\ 0 & 0 & 0\end{bmatrix}\frac{1}{\hat{j}}\begin{bmatrix} s & \ell & j\\ 1 & 1 & 0\\ s & k & j\end{bmatrix}\begin{bmatrix} s & k & j\\ s & k & j\\ 0 & 0 & 0\end{bmatrix}[s|\sigma^{[1]}|s][\ell|L^{[1]}|k\,]. \tag{4.25}$$

Note that we have used the selection rules emanating from the end-boxes to immediately assign the required values to the new angular momenta arising at the exit of the second recoupling box, viz. s and k. The elementary invariant matrix elements associated with the end-boxes are given in eqs.(4.18) and (4.19). We see that the matrix element $[\ell|L^{[1]}|k]$ imposes the selection rule $k = \ell$. This is easy to understand : as the angular momentum operator $L^{[1]}$ is an axial vector it forbids parity change, and from the selection rule $\Delta\ell = 0,\pm1$ only the first possibility survives. Simplifying eq.(4.25) we obtain

$$M = \frac{1}{2}\,\hat{1}\;\frac{1}{\hat{s}\,\hat{\ell}}\begin{bmatrix} s & \ell & j \\ 1 & 1 & 0 \\ s & \ell & j \end{bmatrix}\,[s|\sigma^{[1]}|s]\;[\ell|L^{[1]}|\ell]\;\delta_{\ell k}. \tag{4.26}$$

Here a non-trivial recoupling coefficient of the 6-j type remains. Carrying out this replacement we find from table 1, Chapter 10

$$M = \frac{1}{2}\,(-)^{s+\ell+1+j}\begin{Bmatrix} s & \ell & j \\ \ell & s & 1 \end{Bmatrix}\left[s\middle|\sigma^{[1]}\middle|s\right]\;\left[\ell\middle|L^{[1]}\middle|\ell\right]$$

$$= (-)^{s+\ell+j}\sqrt{\frac{3}{2}}\,\sqrt{\ell(\ell+1)(2\ell+1)}\begin{Bmatrix} s & \ell & j \\ \ell & s & 1 \end{Bmatrix}. \tag{4.27}$$

Generally it is preferable to keep the form with the square 9-j and the elementary invariants, here eq. (4.26), since it permits immediate proofreading and checking from the graph.

The spin-spin interaction

We calculate the matrix element of the spin-spin coupling,

$$\vec{\sigma}_1\cdot\vec{\sigma}_2 = \hat{1}\;[\sigma_1^{[1]}\sigma_2^{[1]}]^{[0]}, \tag{4.28}$$

between the two-body spin state, $s = ½$,

$$\psi_S = \hat{S}\ \left[W^{[S]}[\chi_1^{[s]}\chi_2^{[s]}]^{[S]}\right]^{[0]}. \tag{4.29}$$

We thus must evaluate

$$M = \langle\psi_S|\vec{\sigma}_1\cdot\vec{\sigma}_2|\psi_S\rangle$$

$$= \hat{S}^2\ \hat{1}\ \left[\left[W^{[S]}[\chi_1^{[s]}\chi_2^{[s]}]^{[S]}\right]^{[0]}\Big|[\sigma_1^{[1]}\sigma_2^{[1]}]^{[0]}\ \Big|\left[W^{[S]}[\chi_1^{[s]}\chi_2^{[s]}]^{[S]}\right]^{[0]}\right]. \tag{4.30}$$

The graph fig.4.8 uses the notation of fig.4.2 for the invariant triple product and gives

$$M = \hat{S}^2\ \hat{1}\ \begin{bmatrix} S & S & 0 \\ S & S & 0 \\ 0 & 0 & 0 \end{bmatrix}\ \frac{1}{\hat{S}}\ \begin{bmatrix} s & s & S \\ 1 & 1 & 0 \\ s & s & S \end{bmatrix}\begin{bmatrix} s & s & 1 \\ s & s & 1 \\ 0 & 0 & 0 \end{bmatrix}\ [s|\sigma^{[1]}|s]\ [s|\sigma^{[1]}|s]$$

$$= -\ 6\ \frac{\hat{1}^2}{\hat{S}\ \hat{s}}\ \begin{bmatrix} s & s & S \\ 1 & 1 & 0 \\ s & s & S \end{bmatrix}, \tag{4.31}$$

where we have replaced the invariant matrix elements by their numerical values, eq.(4.18), and the square 9-j coefficients containing more than one zero by their algebraic expressions, eq.(1.15).

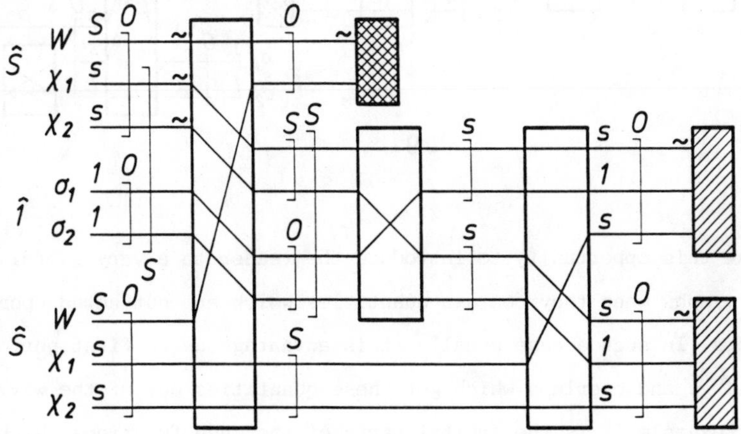

4.8

As another example we evaluate the matrix element of the same spin-spin operator $\vec{\sigma}_1 \cdot \vec{\sigma}_2$ between two-body states in *jj*-coupling,

$$\Psi_I = \hat{I} \left[W^{[I]}[\psi^{[i]}_{s\ell}\psi^{[j]}_{sk}]^{[I]}\right]^{[0]} \equiv \hat{I}\ [W^{[I]}\psi^{[i]}_{s\ell}\psi^{[j]}_{sk}]^{[0]}. \tag{4.32}$$

In order not to have to treat symmetrization, which will be done in Chapter 6, we consider the case where the two particles are non-identical. The complete matrix element is

$$M = \hat{I}^2\ \hat{1}\ \left[[W^{[I]}\psi^{[i]}_{s\ell}\psi^{[j]}_{sk}]^{[0]}\right| [\sigma^{[1]}_1\sigma^{[1]}_2]^{[0]} \left|[W^{[I]}\psi^{[i']}_{s\ell'}\psi^{[j']}_{sk'}]^{[0]}\right]. \tag{4.33}$$

and is represented on fig.4.9.

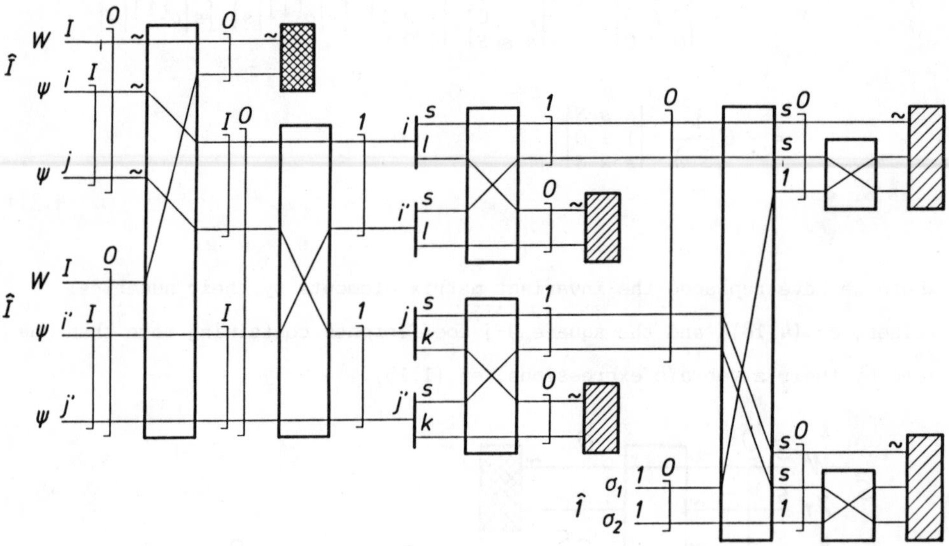

4.9

We take this opportunity to introduce the reader to a very useful tool in drawing graphs when they contain quantities which are not acted upon by the operators. In such a case usually it is advantageous to first perform the recouplings and overlaps which get these quantities out of the way. In our present example it is the orbital parts of the wave functions which are not acted upon by the operators. Furthermore, since the operator $\sigma_1 \cdot \sigma_2$ is an invariant, it is a zero angular momentum quantity and one may insert it anywhere in the graph, provided it is finally properly recoupled into the end-boxes. We use this freedom to insert the operator at the bottom of the graph.We read off, noting $\ell = \ell'$ and $k = k'$ from the end-boxes,

$$M = \hat{I}^2\ \hat{1} \begin{bmatrix} I & I & 0 \\ I & I & 0 \\ 0 & 0 & 0 \end{bmatrix} \frac{1}{\hat{I}} \begin{bmatrix} i & j & I \\ i' & j' & I \\ 1 & 1 & 0 \end{bmatrix} \begin{bmatrix} s & \ell & i \\ s & \ell & i' \\ 1 & 0 & 1 \end{bmatrix} \begin{bmatrix} s & k & j \\ s & k & j' \\ 1 & 0 & 1 \end{bmatrix} \hat{\ell}\ \hat{k} \begin{bmatrix} 1 & 1 & 0 \\ 1 & 1 & 0 \\ 0 & 0 & 0 \end{bmatrix}$$

$$\times\ (-)^{1+s-s}(-)^{1+s-s}\left[s\middle|\sigma^{[1]}\middle|s\right]\ \left[s\middle|\sigma^{[1]}\middle|s\right]$$

$$= -6\,\frac{\hat{\ell}_1\hat{k}_2}{\hat{I}\ \hat{1}} \begin{bmatrix} i & j & I \\ i' & j' & I \\ 1 & 0 & 1 \end{bmatrix} \begin{bmatrix} s & \ell & i \\ s & \ell & i' \\ 1 & 0 & 1 \end{bmatrix} \begin{bmatrix} s & k & j \\ s & k & j' \\ 1 & 0 & 1 \end{bmatrix}, \tag{4.34}$$

with the value $i\sqrt{6}$ for the invariant matrix elements from eq.(4.18), and the values $\hat{\ell}$ and $\hat{k}$ for the invariant norm end-boxes of the orbital overlaps, eq.(2.13).

Two-particle matrix element

As a final example we compute the matrix element of a two particle system acted upon by a factorized two-body operator. The problems of symmetrization and antisymmetrization will be discussed in Chapters 6 and 7. We define

$$\psi_{If}(1,2) = \hat{I}[W_f^{[I]}\ \psi_f^{[I]}(1,2)]^{[0]}, \quad \psi_f^{[I]}(1,2) = [\psi^{[i_1]}(1)\ \psi^{[i_2]}(2)]^{[I]},$$

$$\psi_{Ji}(1,2) = \hat{J}[W_i^{[J]}\ \psi_i^{[J]}(1,2)]^{[0]}, \quad \psi_i^{[J]}(1,2) = [\psi^{[j_1]}(1)\ \psi^{[j_2]}(2)]^{[J]},$$

$$\Omega_\lambda(1,2) = \hat{\lambda}[\omega^{[\lambda]}\ \Omega^{[\lambda]}(1,2)]^{[0]}, \quad \Omega^{[\lambda]}(1,2) = [\Omega^{[\lambda_1]}(1)\ \Omega^{[\lambda_2]}(2)]^{[\lambda]}.$$

We read off fig.4.10 the expression for

$$M = \langle\psi_{If}|\ \Omega_\lambda\ |\psi_{Ji}\rangle$$

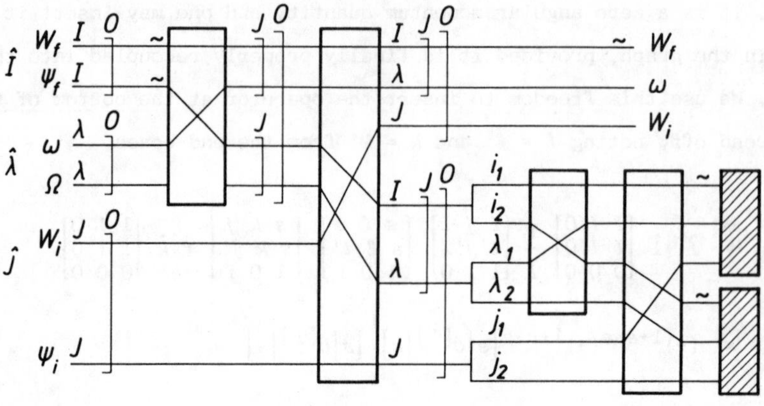

4.10

$$M = \hat{I}\,\hat{\lambda}\,\hat{J} \begin{bmatrix} I & I & 0 \\ \lambda & \lambda & 0 \\ J & J & 0 \end{bmatrix} \begin{bmatrix} J & J & 0 \\ J & J & 0 \\ 0 & 0 & 0 \end{bmatrix} \begin{bmatrix} i_1 & i_2 & I \\ \lambda_1 & \lambda_2 & 0 \\ j_1 & j_2 & J \end{bmatrix} \begin{bmatrix} j_1 & j_2 & J \\ j_1 & j_2 & J \\ 0 & 0 & 0 \end{bmatrix}$$

$$\times \left[\tilde{W}_f^{[I]} \omega^{[\lambda]} W_i^{[J]}\right]^{[0]} \left[\psi^{[i_1]} \middle| \Omega^{[\lambda_1]} \middle| \psi^{[j_1]}\right] \left[\psi^{[i_2]} \middle| \Omega^{[\lambda_2]} \middle| \psi^{[j_2]}\right]$$

$$= \hat{I}\,\hat{\lambda}\,\hat{J} \begin{Bmatrix} i_1 & i_2 & I \\ \lambda_1 & \lambda_2 & \lambda \\ j_1 & j_2 & J \end{Bmatrix} \left[\tilde{W}_f^{[I]} \omega^{[\lambda]} W_i^{[J]}\right]^{[0]} \left[\psi^{[i_1]} \middle| \Omega^{[\lambda_1]} \middle| \psi^{[j_1]}\right] \left[\psi^{[i_2]} \middle| \Omega^{[\lambda_2]} \middle| \psi^{[j_2]}\right] \tag{4.35}$$

In order to exhibit the symmetries we have replaced the square 9-j coefficient by the usual Wigner 9-j, eq.(1.5). We note also the modular character of the graph 4.10 : its first part reproduces fig.4.5 and the structure of eq.(4.10). One may say that the remaining part of fig.4.10 evaluates the invariant matrix element $\left[\psi_f^{[I]} | \Omega^{[\lambda]} | \psi_i^{[J]}\right]$ of eq.(4.11).

This Chapter completes the background information necessary to carry out angular momentum calculations with the Invariant Graph Method. The next Chapters treat specific applications. Chapter 10 will collect the rules for drawing graphs and writing their algebraic expressions, which have been already put to work in these introductory Chapters. Chapter 10 will also give the tables of actual values of the invariant matrix elements which are needed for most applications.

Part two

APPLICATIONS

Chapter 5

VECTOR CALCULUS

Vector calculus is a sub-class of angular momentum recoupling. It is important because many physical quantities are constructed from cartesian vectors, in particular the differential operators. The relation between cartesian vectors and their properly phased spherical tensor components have been given in eqs.(3.13) and (3.31). In this Chapter, after expressing the different products of two vectors in their invariant tensorial form, we show how the graphical method is used to evaluate the action of differential vector operators on both ordinary and vector spherical harmonics. The cases which are presented cover most of the forms of the vector operators encountered in the applications. As an example we apply the method to evaluate a relativistic Dirac spinor.

Products of two vectors

Vector algebra is based on the definition of the scalar, vector and tensor products in terms of the unit vectors $e^{[1]}$, eq.(3.13). For the scalar product, eq.(3.32), we have ($\vec{e}\cdot\vec{e} = e_x^2 + e_y^2 + e_z^2 = 3$)

$$\vec{e}\ .\ \vec{e} = \hat{1}\ [e^{[1]}e^{[1]}]^{[0]} = \hat{1}^2; \tag{5.1}$$

likewise we find

$$\vec{e} \times \vec{e} = \sqrt{2}\,[e^{[1]}e^{[1]}]^{[1]}_{m} = \sqrt{2}\,e^{[1]}_{m}, \tag{5.2}$$

$$\vec{e} \otimes \vec{e} = [e^{[1]}e^{[1]}]^{[2]}_{m} = e^{[2]}_{m}. \tag{5.3}$$

Note the factor $\sqrt{2}$ for the vector product.

The unit vector and unit tensor of course have the invariant norms, eq.(2.13),

$$\left[e^{[1]}\middle|e^{[1]}\right] = \hat{1}, \tag{5.4}$$

$$\left[e^{[2]}\middle|e^{[2]}\right] = \hat{2}. \tag{5.5}$$

A vector $\vec{A}$ can be written in invariant form according to

$$\vec{A} = A_x\vec{e}_x + A_y\vec{e}_y + A_z\vec{e}_z = \hat{1}\,[A^{[1]}e^{[1]}]^{[0]}, \tag{5.6}$$

where the spherical tensor components $A^{[1]}_{m}$ are related to the cartesian amplitudes A_x, A_y, A_z by eq.(3.31). Then the different products of two vectors in their invariant forms are

$$\vec{A} \cdot \vec{B} = \hat{1}^2\,[A^{[1]}e^{[1]}]^{[0]} \cdot [B^{[1]}e^{[1]}]^{[0]}, \tag{5.7}$$

$$\vec{A} \times \vec{B} = \hat{1}^2\,[A^{[1]}e^{[1]}]^{[0]} \times [B^{[1]}e^{[1]}]^{[0]}, \tag{5.8}$$

$$\vec{A} \otimes \vec{B} = \hat{1}^2\,[A^{[1]}e^{[1]}]^{[0]} \otimes [B^{[1]}e^{[1]}]^{[0]}. \tag{5.9}$$

These products can be re-expressed. The graph of fig.5.1 yields the result

$$\hat{1}\,[A^{[1]}e^{[1]}]^{[0]}\;\hat{1}\,[B^{[1]}e^{[1]}]^{[0]} = \sum_{K} \hat{1}^2 \begin{bmatrix} 1 & 1 & 0 \\ 1 & 1 & 0 \\ K & K & 0 \end{bmatrix} \left[[e^{[1]}e^{[1]}]^{[K]}[A^{[1]}B^{[1]}]^{[K]}\right]^{[0]}. \tag{5.10}$$

where for the scalar product $K = 0$, for the vector product $K = 1$, and for the tensor product $K = 2$. We have from eqs.(5.1), (5.2) and (5.3) respectively

$$\vec{A} \cdot \vec{B} = \hat{1}\ [A^{[1]}B^{[1]}]^{[0]}, \tag{5.11}$$

$$\vec{A} \times \vec{B} = \sqrt{2}\ \hat{1}\ [A^{[1]}B^{[1]}e^{[1]}]^{[0]}, \tag{5.12}$$

$$\vec{A} \otimes \vec{B} = \hat{2}\ [A^{[1]}B^{[1]}e^{[2]}]^{[0]}. \tag{5.13}$$

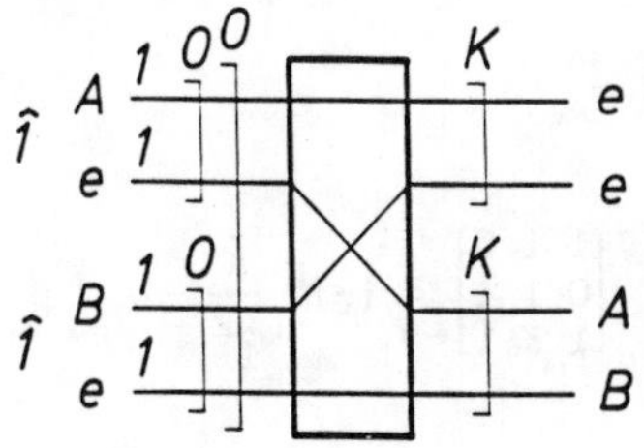

5.1

A special case arises when one of the vector, say $\vec{B}$, is a vector spherical harmonic function,

$$\vec{Y}_{J\ell} = \hat{J}\ [W^{[J]}e^{[1]}Y^{[\ell]}]^{[0]}. \tag{5.14}$$

We calculate its different products with a vector $\vec{A}$, eq.(5.6), using eqs.(5.11) through (5.13). They are given by the graph of fig. 5.2. The crossing box yields the phase (+) and we get the general expression

$$\hat{1}\ [A^{[1]}e^{[1]}]^{[0]}\ \hat{J}\ [W^{[J]}e^{[1]}Y^{[\ell]}]^{[0]} =$$

$$\sum_K \hat{1}\ \hat{J} \begin{bmatrix} 1 & 1 & 0 \\ 0 & 1 & 1 \\ 1 & K & 1 \end{bmatrix} \left[W^{[J]}\left[A^{[1]}[e^{[1]}e^{[1]}]^{[K]}\right]^{[1]}Y^{[\ell]}\right]^{[0]}. \tag{5.15}$$

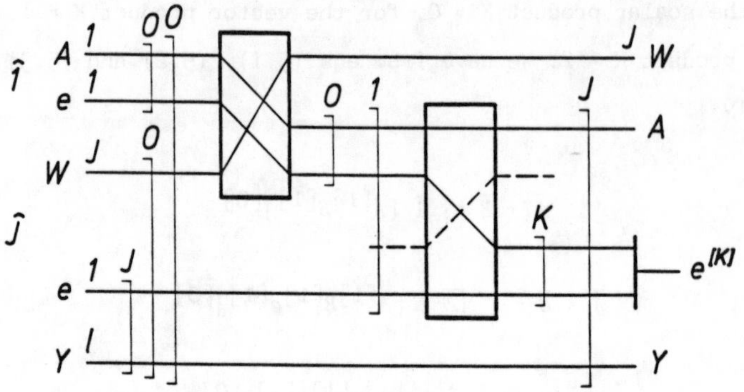

5.2

Since

$$\begin{bmatrix} 1 & 1 & 0 \\ 0 & 1 & 1 \\ 1 & K & 1 \end{bmatrix} = (-)^K \frac{\hat{K}}{\hat{1}^2}, \tag{5.16}$$

the scalar product, $K = 0$, is

$$\vec{A} \cdot \vec{Y}_{J\ell} = \hat{J}\, [W^{[J]} A^{[1]} Y^{[\ell]}]^{[0]}, \tag{5.17}$$

the vector product, $K = 1$, is (the factor $(-)^K$ provides for the change of coupling order between $A^{[1]}$ and $e^{[K]}$ in the expression)

$$\vec{A} \times \vec{Y}_{J\ell} = \sqrt{2}\, \hat{J}\, \left[W^{[J]} [e^{[1]} A^{[1]}]^{[1]} Y^{[\ell]}\right]^{[0]}, \tag{5.18}$$

and the tensor product, $K = 2$, is

$$\vec{A} \otimes \vec{Y}_{J\ell} = \frac{\hat{J}}{\hat{1}}\, \hat{2}\, \left[W^{[J]} [e^{[2]} A^{[1]}]^{[1]} Y^{[\ell]}\right]^{[0]}. \tag{5.19}$$

Differential operators : grad, div, curl

The formulae of the previous section can be applied to vector analysis by specifying for $\vec{A}$ the nabla operator $\vec{\nabla}$ which in invariant form is

$$\vec{\nabla} = \vec{e}_x \nabla_x + \vec{e}_y \nabla_y + \vec{e}_z \nabla_z = \hat{1}\ [e^{[1]}\nabla^{[1]}]^{[0]}. \tag{5.20}$$

For a scalar function f the gradient operation is written as

$$grad\ f = \vec{\nabla} f = \hat{1}\ [e^{[1]}\nabla^{[1]}]^{[0]} f = \hat{1}\ [e^{[1]}(\nabla^{[1]} f)]^{[0]}. \tag{5.21}$$

The *div* and *curl* operations applied to a vector field $\vec{B}$ are written from eqs.(5.11) and (5.12) respectively as

$$div\ \vec{B} = \vec{\nabla}\ .\ \vec{B} = \hat{1}\ [\nabla^{[1]} B^{[1]}]^{[0]}, \tag{5.22}$$

$$curl\ \vec{B} = \vec{\nabla} \times \vec{B} = \sqrt{2}\ \ \hat{1}\ [e^{[1]}\nabla^{[1]} B^{[1]}]^{[0]}. \tag{5.23}$$

We now apply these formulae to normalized scalar orbital functions

$$\varphi^{[\ell]}_{\alpha m}(\vec{r}) = F_{\alpha\ell}(r)\ Y^{[\ell]}_m(\hat{r}), \tag{5.24}$$

and to normalized vector spherical functions defined as

$$\vec{\psi}^{[J]}_{\alpha\ell M}(\vec{r}) = F_{\alpha\ell}(r)\ \vec{Y}^{[J]}_{\ell M}(\hat{r}) = [e^{[1]}\varphi^{[\ell]}_{\alpha}(\vec{r})]^{[J]}_M. \tag{5.25}$$

or in invariant form

$$\vec{\psi}_{\alpha\ell J} = \hat{J}\ [W^{[J]}\ \vec{\psi}^{[J]}_{\alpha\ell}]^{[0]} = \hat{J}\ \left[W^{[J]}[e^{[1]}\varphi^{[\ell]}_{\alpha}]^{[J]}\right]^{[0]}. \tag{5.26}$$

The matrix element of the gradient operator $\vec{\nabla}$ between two orbital functions

$$M = \langle \varphi_{\alpha k} |\ \vec{\nabla}\ | \varphi_{\beta\ell} \rangle$$

$$= \hat{k}\ \hat{1}\ \hat{\ell}\ \left[[W^{[k]}\varphi^{[k]}_{\alpha}]^{[0]} |\ [e^{[1]}\nabla^{[1]}]^{[0]}\ | [W^{[\ell]}\varphi^{[\ell]}_{\beta}]^{[0]}\right] \tag{5.27}$$

is evaluated with the graph 5.3 which yields

$$M = \hat{k}\,\hat{\imath}\,\hat{\ell} \begin{bmatrix} k & k & 0 \\ 1 & 1 & 0 \\ \ell & \ell & 0 \end{bmatrix} \begin{bmatrix} \ell & \ell & 0 \\ \ell & \ell & 0 \\ 0 & 0 & 0 \end{bmatrix} [\tilde{W}^{[k]} e^{[1]} W^{[\ell]}]^{[0]} \left[\varphi_\alpha^{[k]} \middle| \nabla^{[1]} \middle| \varphi_\beta^{[\ell]}\right]$$

$$= \left[\varphi_\alpha^{[k]} | \nabla^{[1]} | \varphi_\beta^{[\ell]}\right] [\rho_{\ell k}^{[1]} e^{[1]}]^{[0]}. \tag{5.28}$$

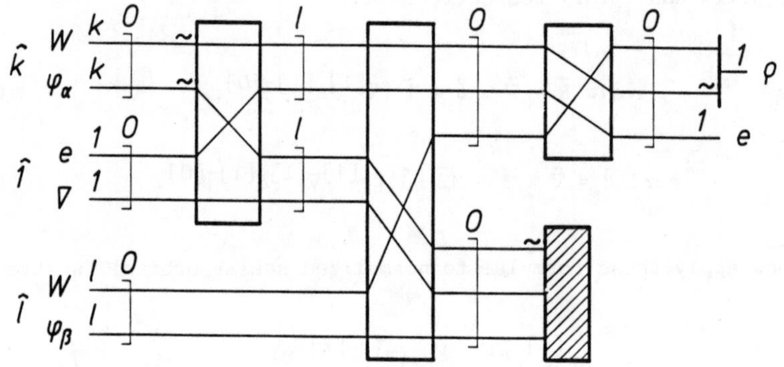

5.3

In the expression (5.28) we have introduced explicitly the coupled quantity $\rho_{\ell k}^{[1]}$, the density matrix, defined by

$$\rho_{\ell k}^{[1]} = [W^{[\ell]} \tilde{W}^{[k]}]^{[1]}. \tag{5.29}$$

It represents in fact the direction of the resulting gradient vector. Thus in eq.(5.28) we have met our first example of the description of the preparation of a system in terms of the density matrix, here $\rho_{\ell k}^{[1]}$. The invariant matrix element of $\nabla^{[1]}$ in eq.(5.28) contains the parity and angular momentum selection rules, $k = \ell \pm 1$. The explicit values of that matrix element are given in Chapter 10.

The matrix element of the *div* operator on a vector spherical multipole $\vec{\psi}_{\beta\ell J}$, eq.(5.26), is as simply obtained noting that the final state is a scalar wave function $\varphi_\alpha^{[k]}$, eq.(5.24). This matrix element writes

$$M = \langle \varphi_{\alpha k} |\ div\ | \vec{\psi}_{\beta \ell J} \rangle$$

$$= \hat{k}\ \hat{1}\ \hat{J} \left[[W^{[k]} \varphi_{\alpha}^{[k]}]^{[0]} \middle| \ [e^{[1]} \nabla^{[1]}]^{[0]} \ \middle| [W^{[J]}\ \vec{\psi}_{\beta \ell}^{[J]}]^{[0]} \right].$$

(5.30)

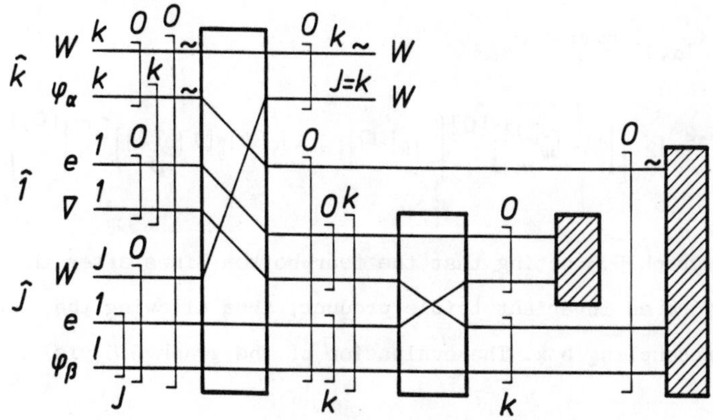

5.4

From the graph 5.4 we note the selection rule on the intermediate angular momentum $J=k$ and we read off the result

$$M = \hat{k}^2\ \hat{1} \begin{bmatrix} k & k & 0 \\ k & k & 0 \\ 0 & 0 & 0 \end{bmatrix} \begin{bmatrix} 1 & 1 & 0 \\ 1 & \ell & k \\ 0 & k & k \end{bmatrix} \hat{1}\ [\tilde{W}^{[k]} W^{[k]}]^{[0]} \left[\varphi_{\alpha}^{[k]} \middle| \nabla^{[1]} \middle| \varphi_{\beta}^{[\ell]} \right]$$

$$= [\tilde{W}^{[k]} W^{[k]}]^{[0]} \left[\varphi_{\alpha}^{[k]} \middle| \nabla^{[1]} \middle| \varphi_{\beta}^{[\ell]} \right].$$

(5.31)

The values of the invariant matrix elements of $\nabla^{[1]}$ are given at the end of Chapter 10.

We now compute the matrix element of the *curl* operator between initial and final vector spherical harmonic functions, eq.(5.26). The *curl* is given by the expression (5.23) ; note the replacement of $\hat{1}$ by $\hat{J}$, the total angular momentum

$$\vec{\nabla}\times\ \vec{\psi}_{\beta\ell J} = \sqrt{2}\ \hat{J}\ \left[W^{[J]}\left[[e^{[1]}\nabla^{[1]}]^{[1]}\ \varphi_{\beta}^{[\ell]}\right]^{[J]}\right]^{[0]} . \tag{5.32}$$

Thus the matrix element is

$$\begin{aligned} M &= \langle\vec{\psi}_{\alpha kJ}\mid curl \mid \vec{\psi}_{\beta\ell J}\rangle \\ &= \hat{J}^2\ \sqrt{2}\ \left[\left[W^{[J]}\vec{\psi}_{\alpha k}^{[J]}\right]^{[0]}\middle|\ \left[W^{[J]}\left[[e^{[1]}\nabla^{[1]}]^{[1]}\varphi_{\beta}^{[\ell]}\right]^{[J]}\right]^{[0]}\right]. \end{aligned} \tag{5.33}$$

We draw the graph 5.5 noting that the four bottom lines after the first recoupling form an invariant triple product, thus allowing the geometry of the second recoupling box. The evaluation of the graph 5.5 yields

$$M = \hat{J}^2\ \sqrt{2}\ \begin{bmatrix} J & J & 0 \\ J & J & 0 \\ 0 & 0 & 0 \end{bmatrix} \begin{bmatrix} 1 & k & J \\ 1 & 1 & 1 \\ 0 & \ell & \ell \end{bmatrix}\ [\tilde{W}^{[J]}W^{[J]}]^{[0]}\ [e^{[1]}|e^{[1]}]\left[\varphi_{\alpha}^{[k]}\middle|\nabla^{[1]}\middle|\varphi_{\beta}^{[\ell]}\right], \tag{5.34}$$

and after simplifications, in particular eq.(5.4), we get for the matrix element of the *curl* operator

$$M = \sqrt{2}\ \hat{1}\ \begin{bmatrix} 1 & k & J \\ 1 & 1 & 1 \\ 0 & \ell & \ell \end{bmatrix}\ [\tilde{W}^{[J]}W^{[J]}]^{[0]}\ \left[\varphi_{\alpha}^{[k]}\middle|\nabla^{[1]}\middle|\varphi_{\beta}^{[\ell]}\right]. \tag{5.35}$$

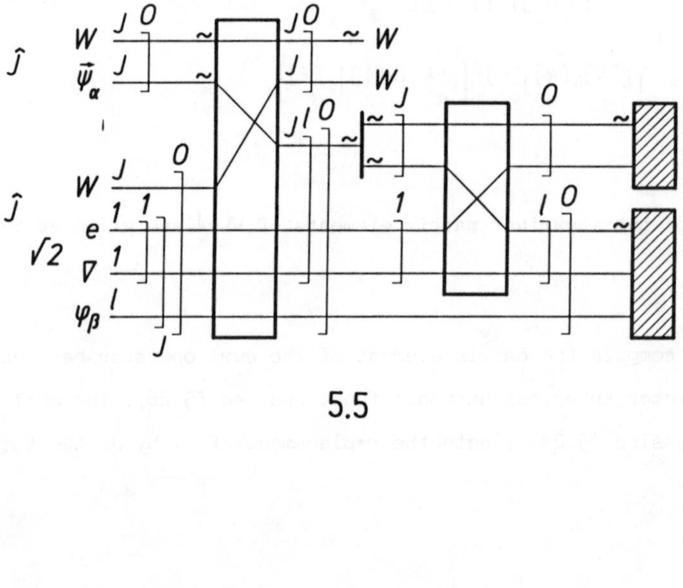

5.5

Action of the spin-momentum operator

Until now we have been concerned with the evaluation of matrix elements. We here show that the action of an operator on a wave function can be reduced to the evaluation of a matrix element by introducing the unit operator eq.(2.24). As an example we consider the computation of the lower component of a Dirac spinor.

A Dirac spinor for the positive energy solutions is written in invariant form as

$$\psi_{\alpha\ell j}(\vec{x}) = \hat{j}\ \left[W^{[j]}\begin{pmatrix} 1 \\ \dfrac{\hat{1}\ [\sigma^{[1]}p^{[1]}]^{[0]}}{E+M}\end{pmatrix}\psi^{[j]}_{\alpha\ell}(\vec{x})\right]^{[0]} = \begin{pmatrix}\psi^{U} \\ \psi^{L}\end{pmatrix}, \tag{5.36}$$

where (s = ½)

$$\psi^{[j]}_{\alpha\ell}(\vec{x}) = [\chi^{[s]}\varphi^{[\ell]}_{\alpha}(\vec{x})]^{[j]}. \tag{5.37}$$

The problem consists in evaluating the lower component ψ^{L}. This is done on the graph 5.6 where use is made of the insertion operation

$$\delta^{3}(\vec{x}-\vec{y}) = \sum_{\beta\lambda j}\hat{j}\left[\tilde{\psi}^{[j]}_{\beta\lambda}(\vec{x})\psi^{[j]}_{\beta\lambda}(\vec{y})\right]^{[0]}. \tag{5.38}$$

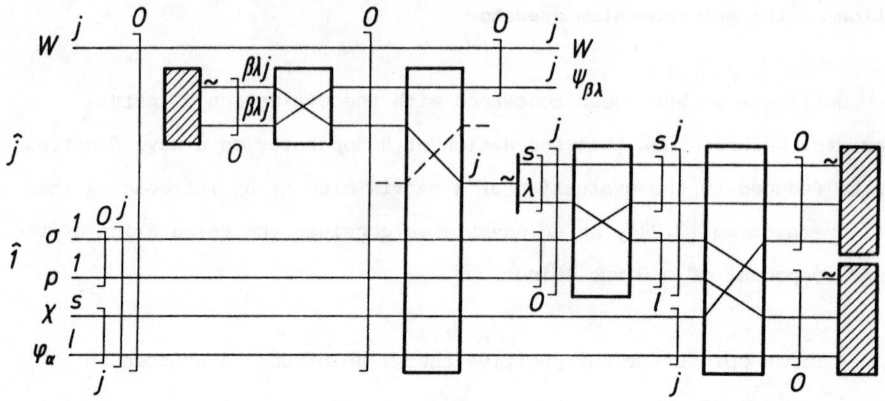

5.6

We read off the result, as always inserting the selection rules beginning from the right of the graph

$$\psi^L = \frac{\hat{j}\ \hat{1}}{E+M}\ \hat{j}\ (-)^{2j}\begin{bmatrix} j & j & 0 \\ 0 & j & j \\ j & 0 & j \end{bmatrix}\begin{bmatrix} s & \lambda & j \\ 1 & 1 & 0 \\ s & \ell & j \end{bmatrix}\begin{bmatrix} s & \ell & j \\ s & \ell & j \\ 0 & 0 & 0 \end{bmatrix}$$

$$\times \sum_\beta\ [W^{[j]}\psi^{[j]}_{\beta\lambda}]^{[0]}\ \left[s\middle|\sigma^{[1]}\middle|s\right]\ \left[\varphi^{[\lambda]}_\beta\middle|p^{[1]}\middle|\varphi^{[\ell]}_\alpha\right]$$

$$= -\frac{1}{E+M}\frac{\hat{j}\ \hat{1}}{\hat{s}\ \hat{\ell}}\ i\ \sqrt{6}\begin{bmatrix} s & \lambda & j \\ 1 & 1 & 0 \\ s & \ell & j \end{bmatrix}\sum_\beta \left[\varphi^{[\lambda]}_\beta\middle|p^{[1]}\middle|\varphi^{[\ell]}_\alpha\right]\ [W^{[j]}\psi^{[j]}_{\beta\lambda}]^{[0]}, \tag{5.39}$$

where the selection rules impose $\lambda=\ell+1$ if $j=\ell+½$ or $\lambda=\ell-1$ if $j=\ell-½$. Introducing the explicit values

$$\begin{bmatrix} s & \lambda & j \\ 1 & 1 & 0 \\ s & \ell & j \end{bmatrix} = \frac{\hat{s}\ \hat{\ell}}{\hat{1}\ \sqrt{6}} \times \begin{cases} \dfrac{1}{\sqrt{\lambda}} & if \quad \lambda = \ell+1, \\ -\dfrac{1}{\sqrt{\ell}} & if \quad \lambda = \ell-1, \end{cases} \tag{5.40}$$

we get the final result

$$\psi^{L} = - i \frac{\hat{j}}{E+M} \sum_{\beta} \left[\varphi^{[\lambda]}_{\beta} \middle| p^{[1]} \middle| \varphi^{[\ell]}_{\alpha}\right] [W^{[j]} \psi^{[j]}_{\beta\lambda}]^{[0]}. \tag{5.41}$$

The expressions for the invariant matrix elements of $p^{[1]}$ are given in Chapter 10.

In the important case of a Bessel radial function, which arises in the multipole expansion of a plane wave spinor field

$$\varphi^{[\ell]}_{pm}(\vec{r}) = \sqrt{\frac{2}{\pi}}\, j_{\ell}(pr)\; Y^{[\ell]}_{m}(\hat{r}), \tag{5.42}$$

the invariant matrix elements of the momentum in eq.(5.41) may be replaced by their values given in Chapter 10, with $p^{[1]} = -i\nabla^{[1]}$. We thus get for the lower spinor component

$$\psi^{L} = - \frac{p}{E+M}\, \hat{j}\, [W^{[j]}\, \psi^{[j]}_{p\lambda}]^{[0]}, \tag{5.43}$$

with $\lambda = \ell+1$ for $j = \ell+½$ or $\lambda = \ell-1$ for $j = \ell-½$.

Chapter 6

MANY-BODY PROBLEM

We give here the complements needed to apply the graphical method to the many-body problem. To this end we only consider the problem of identical particles in a single shell.

The calculation of many-body matrix elements requires to perform simultaneously the angular momentum coupling and the particle symmetrization (for bosons) or antisymmetrization (for fermions). For a large number of particles the evaluation is best performed by means of the fractional parentage coefficients. When using such expansions, the CFP coefficients can be represented in the invariant graphs with appropriate symbols. This way all the features of the method can be extended to the many-body problem. In the case of few particles, the antisymmetrization or symmetrization can be carried through explicitly, as we shall show for 3- and 4-body systems. These results yield the starting point for the recursive computation of the CFP's.

Fractional parentage coefficients

We first define the symmetrized (antisymmetrized) and normalized two-particle wave functions in a single j shell. They are the starting point of the procedure. They are of the form

$$\Psi_I(1,2) = \hat{I}\,\frac{1}{2}\left[W^{[I]}\Big([\varphi^{[j]}(1)\varphi^{[j]}(2)]^{[I]} \pm [\varphi^{[j]}(2)\varphi^{[j]}(1)]^{[I]}\Big)\right]^{[0]}. \quad (6.1)$$

It is easily verified that the norm has four contributions yielding 1 for I even and 0 for I odd. Thus we as well can use the simpler form

$$\Psi_I(1,2) = \hat{I}\left[W^{[I]}[\varphi^{[j]}(1)\varphi^{[j]}(2)]^{[I]}\right]^{[0]} = \hat{I}\,[W^{[I]}\Psi^{[I]}(1,2)]^{[0]}, \quad (6.2)$$

where the particle indices are written in ascending order, and with the prescription I even. We will adopt this procedure from now on. We have for the invariant norm the usual value, eq.(2.13), I even

$$\left[\Psi^{[I]}(1,2)\middle|\Psi^{[I]}(1,2)\right] = \hat{I}. \quad (6.3)$$

We now consider n identical particles in a shell j. We define the properly symmetrized or antisymmetrized but non-orthonormal states labelled by the total angular momentum I and by an index α which stands for all the other quantum numbers needed to fully specify the state

$$\begin{aligned}\Phi_{\alpha I}(1,2,\ldots n) &= \hat{I}\,[W^{[I]}\Phi_\alpha^{[I]}(1,2,\ldots n)]^{[0]} \\ &= \hat{I}\sum_{\beta K}\mathcal{A}\left[W^{[I]}[\Psi_\beta^{[K]}(1,2,\ldots n\text{-}1)\varphi^{[j]}(n)]^{[I]}\right]^{[0]}.\end{aligned} \quad (6.4)$$

Here $\Psi_\beta^{[K]}(1,2,\ldots n\text{-}1)$ is a complete set of wave functions for n-1 particles which is not only symmetrized (antisymmetrized) in the particle indices 1,2...n-1 but also normalized. $\mathcal{A}$ is the symmetrization (antisymmetrization) operator for the last particle (index n) and the symmetrized (antisymetrized) particle set (1,2,...n-1). Non-orthonormal symmetrized (antisymmetrized) states are denoted by Φ to distinguish them from the orthonormal sets Ψ.

We now proceed to the properly symmetrized (antisymmetrized) and normalized states for n particles. They are written by means of expansions which define either one-particle fractional parentage coefficients (CFP's)

$$\Psi^{[I]}_{\alpha M}(1,2,\ldots n) = \sum_{\beta K} (j^{n-1}\beta K;\ j|\}j^n\alpha I)\ [\Psi^{[K]}_{\beta}(1,2,\ldots n\text{-}1)\varphi^{[j]}(n)]^{[I]}_{M}, \quad (6.5)$$

or two-particle CFP's

$$\Psi^{[I]}_{\alpha M}(1,2,\ldots n) = \sum_{\beta L} (j^{n-2}\beta K;\ j^2L\|\}j^n\alpha I)$$

$$\times \left[\Psi^{[K]}_{\beta}(1,2,\ldots n\text{-}2)[\varphi^{[j]}(n\text{-}1)\varphi^{[j]}(n)]^{[L]}\right]^{[I]}_{M}. \quad (6.6)$$

In this equation the two coupled particles, written in the order $\varphi^{[j]}(n\text{-}1)\varphi^{[j]}(n)$, have the proper symmetry by imposing L even. Their coupled wave function is normalized, as explained above, eq.(6.2). The quantum numbers β in eqs. (6.5) and (6.6) define complete sets with angular momentum K in the n-1 and n-2 particle spaces respectively.

For representing the one- and two-particle CFP's respectively in the recoupling graphs we introduce the symbols of fig.6.1 for $(j^{n-1}\beta K;\ j|\}j^n\alpha I)$ and fig.6.2 for $(j^{n-2}\beta K;\ j^2L\|\}j^n\alpha I)$.

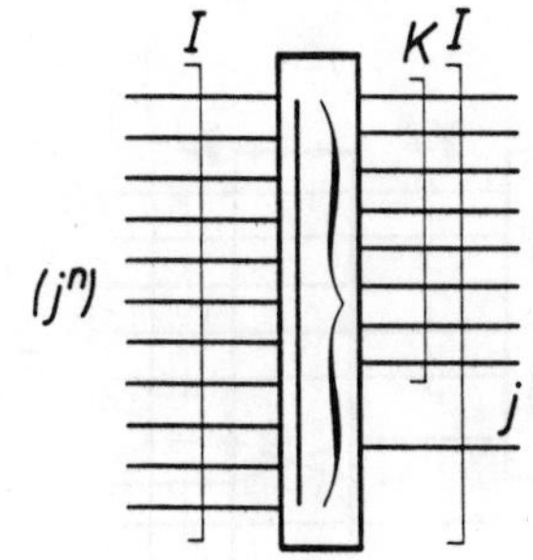

6.1

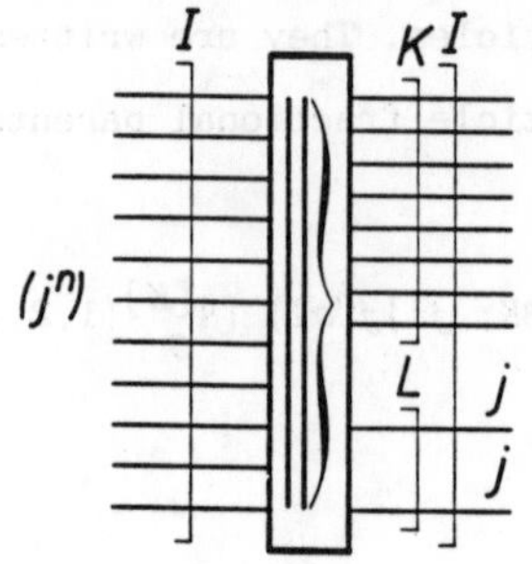

6.2

A method for evaluating recursively the CFP's is given at the end of this Chapter. The CFP's are real. The normalization of the states entails the orthognality relations

$$\sum_{\beta K} (j^{n-1}\beta K;j|\}j^n\alpha I)\ (j^{n-1}\beta K;j|\}j^n\alpha' I') = \delta_{\alpha\alpha'}\delta_{II'}, \tag{6.7}$$

$$\sum_{\beta KL} (j^{n-2}\beta K;j^2L||\}j^n\alpha I)\ (j^{n-2}\beta K;j^2L||\}j^n\alpha' I') = \delta_{\alpha\alpha'}\delta_{II'}. \tag{6.8}$$

As an example of the use of the graphical representation of the CFP's, we compute the relation between one- and two-particle CFP's. It is given by the graph 6.3 which reads

$$\sum_{\beta K} (j^{n-1}\beta K;\ j|\}j^n\alpha I)(j^{n-2}\gamma J;\ j|\}j^{n-1}\beta K) \begin{bmatrix} J & j & K \\ 0 & j & j \\ J & L & I \end{bmatrix} = (j^{n-2}\gamma J;\ j^2L||\}j^n\alpha I). \tag{6.9}$$

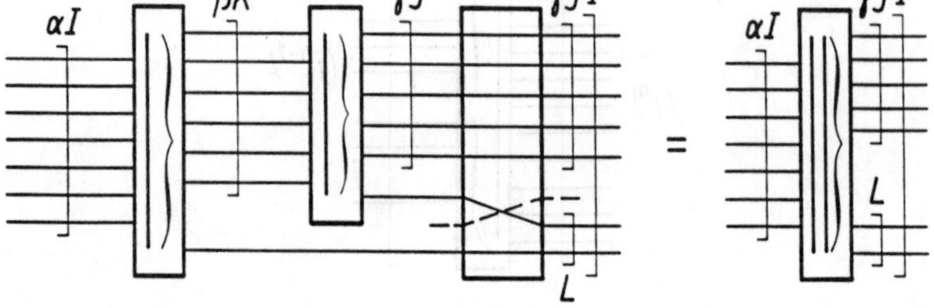

6.3

Many-body matrix elements

A one-body operator acting on an n-particle system is symmetrized relatively to the particle indices i. We of course use its invariant form (3.26)

$$\Omega_\lambda = \sum_{i=1}^{n} \hat{\lambda}\, [\omega^{[\lambda]}\Omega^{[\lambda]}(i)]^{[0]}. \tag{6.10}$$

The expression for its matrix element is

$$M = \langle\Psi_{\alpha I}|\Omega_\lambda|\Psi_{\beta J}\rangle$$

$$= n\, \hat{I}\hat{\lambda}\hat{J}\Big[[W^{[I]}\Psi^{[I]}_\alpha(1..n)]^{[0]}\Big|[\omega^{[\lambda]}\Omega^{[\lambda]}(n)]^{[0]}\Big|[W^{[J]}\Psi^{[J]}_\beta(1..n)]^{[0]}\Big]. \tag{6.11}$$

Note that for convenience we have singled out particle n in the one-body operator. This is allowed in view of the symmetrized (antisymmetrized) wave functions ; hence the factor n in front of the right-hand side expression. The matrix element (6.11) is evaluated on the graph of fig.6.4, which yields after trivial simplifications

$$M = n(-)^\lambda \sum_{\gamma K} \hat{K}\, (j^{n-1}\gamma K;\, j|\}j^n\alpha I)(j^{n-1}\gamma K;\, j|\}j^n\beta J) \begin{bmatrix} K & j & I \\ K & j & J \\ 0 & \lambda & \lambda \end{bmatrix}$$

$$[\tilde{W}^{[I]}W^{[J]}\omega^{[\lambda]}]^{[0]}\ \big[\varphi^{[j]}(1)\big|\Omega^{[\lambda]}(1)\big|\varphi^{[j]}(1)\big]. \tag{6.12}$$

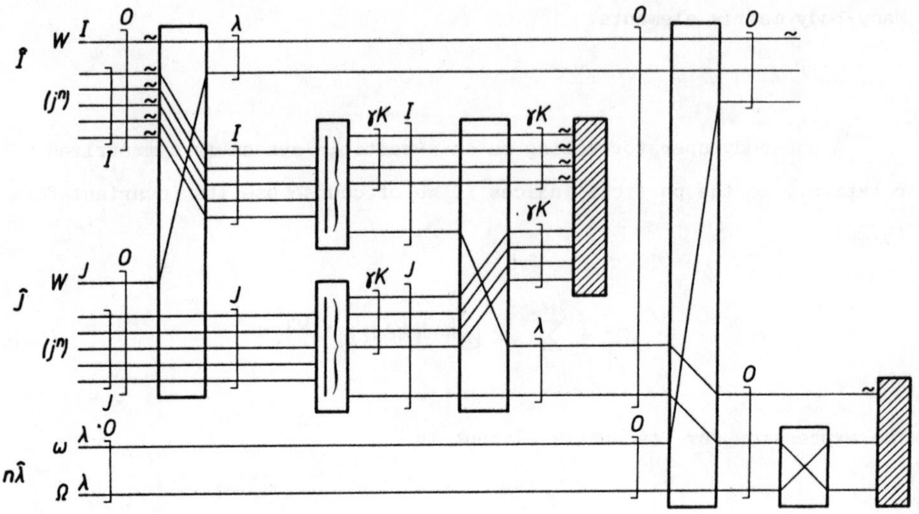

6.4

In drawing the graph 6.4 we have used the selection rule $\gamma = \gamma'$ arising from the overlap of the spectator particles. By the definition of CFP's this overlap is the invariant norm of normalized wave functions ; hence its value $\hat{K}\, \delta_{\gamma\gamma'}$.

Likewise the two-body operator acting on an n-body system, symmetrized relatively to the particle indices i, j

$$\Omega_\lambda = \frac{1}{2} \sum_{i \neq j=1}^{n} \hat{\lambda}\, [\omega^{[\lambda]}\Omega^{[\lambda]}(i,j)]^{[0]}, \tag{6.13}$$

leads to the matrix element, singling out particles n-1 and n

$$M = \frac{n(n-1)}{2}\, \hat{I}\hat{\lambda}\hat{J} \left[[W^{[I]}\Psi_\alpha^{[I]}(1..n)]^{[0]} \middle| [\omega^{[\lambda]}\Omega^{[\lambda]}(n-1,n)]^{[0]} \middle| [W^{[J]}\Psi_\beta^{[J]}(1..n)]^{[0]} \right] \tag{6.14}$$

It is evaluated by reading the graph 6.5 which yields

$$M = \frac{n(n-1)}{2} \sum_{\gamma K} \sum_{LL'} (-)^{\lambda+L'-L} \hat{K} \, (j^{n-2}\gamma K;\ j^2 L \| \} j^n \alpha I)(j^{n-2}\gamma K;\ j^2 L' \| \} j^n \beta J)$$

$$\times \begin{bmatrix} K & L & I \\ K & L' & J \\ 0 & \lambda & \lambda \end{bmatrix} [\tilde{W}^{[I]} W^{[J]}{}_{\omega}{}^{[\lambda]}]^{[0]} \Big[[\varphi^{[j]}(1)\varphi^{[j]}(2)]^{[L]} \Big| \Omega^{[\lambda]}(1,2) \Big| [\varphi^{[j]}(1)\varphi^{[j]}(2)]^{[L']} \Big]. \tag{6.15}$$

Of course in this matrix element the total angular momenta L and L' of the two particles in a same shell j are even, eq.(6.2), to ensure symmetrized (antisymmetrized) and orthonormalized states.

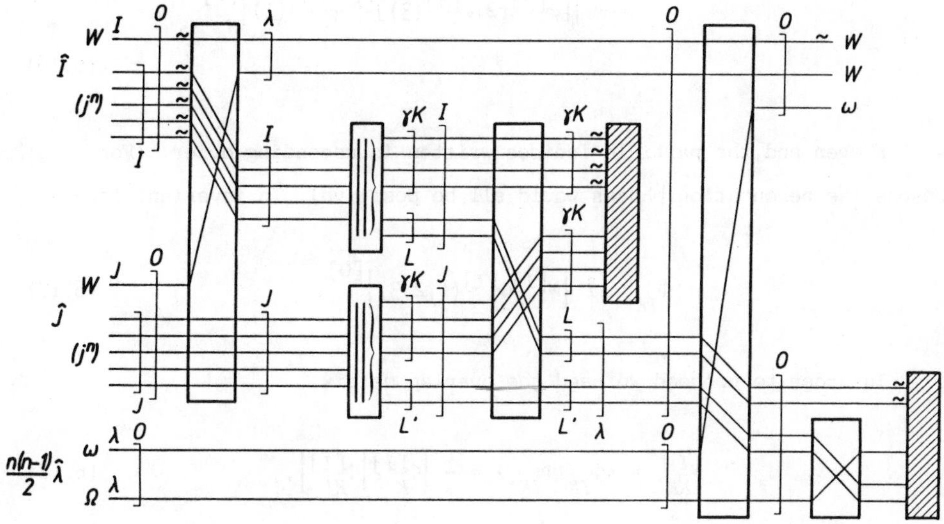

6.5

Evaluation of CFP's

We now construct explicitly with the Invariant Graph method normalized, orthogonal, symmetrized (antisymmetrized) complete sets of wave functions with good angular momentum for systems made of 3 and 4 identical particles in a shell j. The solution to this problem will give us a method for computing recursively the CFP's for n-particle systems.

Three particles

Let us consider the case of fermions. We define a set of non-normalized, non-orthogonal but antisymmetrized wave functions

$$\Phi^{[I]}_{KM}(1,2,3) = \left[[\varphi^{[j]}(1)\varphi^{[j]}(2)]^{[K]}\varphi^{[j]}(3)\right]^{[I]}_M - \left[[\varphi^{[j]}(1)\varphi^{[j]}(3)]^{[K]}\varphi^{[j]}(2)\right]^{[I]}_M - \left[[\varphi^{[j]}(2)\varphi^{[j]}(3)]^{[K]}\varphi^{[j]}(1)\right]^{[I]}_M \tag{6.16}$$

with K even and the particle indices written in ascending order. (For bosons the permutation phases would all be positive). In invariant form

$$\Phi_{IK} = \hat{I}\left[W^{[I]}\Phi^{[I]}_K(1,2,3)\right]^{[0]}. \tag{6.17}$$

In order to proceed we need the overlap matrix

$$N^I_{KK'} = \langle\Phi_{IK}|\Phi_{IK'}\rangle = \frac{1}{\hat{I}}\left[\Phi^{[I]}_K\middle|\Phi^{[I]}_{K'}\right], \tag{6.18}$$

which is non-diagonal in K. Its matrix elements are sums of 9 terms, of which in fig.6.6 we draw the three which correspond to permutations in the ket. The six other diagrams, not drawn, are generated by particle index permutation in the bra and yield the same contributions as above ; hence a factor 3. We have separated out at the left top of the graph the common part which overlaps out the amplitudes $W^{[I]}$. This part yields a factor $1/\hat{I}$. In each graph the top end-box is the invariant norm eq.(6.3) of the antisymmetrized two-body state, which has the value $\hat{K}$. The result is

$$N^I_{KK'} = 3\left(\delta_{KK'} - 2\begin{bmatrix} j & j & K' \\ j & 0 & j \\ K & j & I \end{bmatrix}\right). \tag{6.19}$$

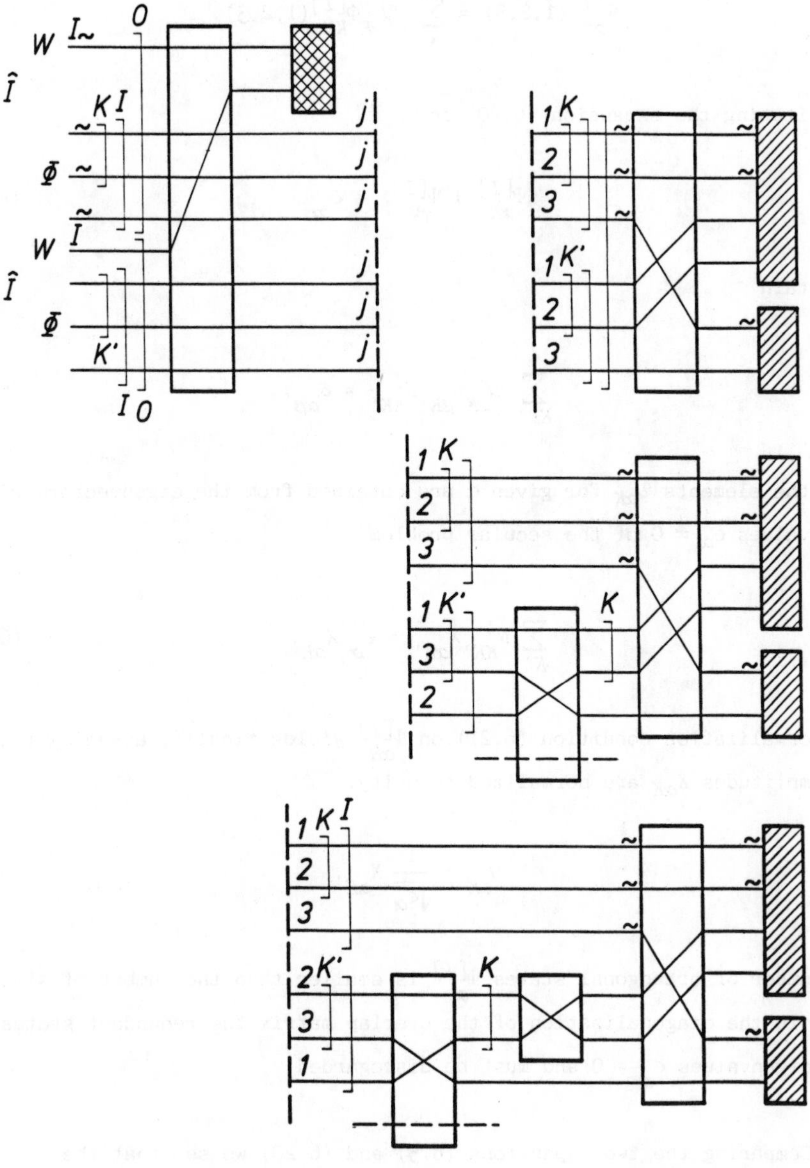

6.6

We now introduce a transformation $\mathcal{U}_{\alpha K}$ which defines the expansion of the unknown orthonormal set $\Psi^{[I]}_{\alpha}(1,2,3)$ in terms of the $\Phi^{[I]}_{K}(1,2,3)$ of eq.(6.16)

$$\Psi^{[I]}_{\alpha M}(1,2,3) = \sum_{K} \mathcal{U}_{\alpha K}\Phi^{[I]}_{KM}(1,2,3). \tag{6.20}$$

Substituting the expansion (6.20) in

$$\langle\Psi^{[I]}_{\alpha M} \mid \Psi^{[I]}_{\beta M}\rangle = \delta_{\alpha\beta}, \tag{6.21}$$

we obtain

$$\sum_{KK'} \mathcal{U}_{\alpha K}\mathcal{U}_{\beta K'}N^{I}_{KK'} = \delta_{\alpha\beta}. \tag{6.22}$$

Thus the elements $\mathcal{U}_{\alpha K}$ for given α are obtained from the eigenvectors with eigenvalues $\epsilon_{\alpha} \neq 0$ of the secular problem

$$\sum_{K'} N^{I}_{KK'}X_{\alpha K'} = \epsilon_{\alpha} X_{\alpha K}. \tag{6.23}$$

The normalization condition (6.21) on $\Psi^{[I]}_{\alpha M}$ yields finally, assuming that the amplitudes $X_{\alpha K}$ are normalized to unity,

$$\mathcal{U}_{\alpha K} = \frac{1}{\sqrt{\epsilon_{\alpha}}} X_{\alpha K}. \tag{6.24}$$

The number of orthogonal states $\Psi^{[I]}_{\alpha}$ is smaller than the number of states $\Phi^{[I]}_{K}$. In the diagonalization of the overlap matrix the redundant states have eigenvalues $\epsilon_{\alpha} = 0$ and must be disregarded.

Comparing the two expansions (6.5) and (6.20) we see that the one-particle CFP's for the 3-fermion system are given by the result (6.24)

$$(j^2K;\ j|\}j^3\alpha I) = \mathcal{U}_{\alpha K}. \tag{6.25}$$

Four particles

Again considering fermions, we follow the same steps, using first the complete set of orthonormal and antisymmetrized 3-particle wave functions $\Psi^{[I]}_{\beta M}$ just obtained above. The 4-particle states, properly antisymmetrized but non-orthonormalized, are given by

$$\begin{aligned}\Phi^{[I]}_{\beta JM}(1,2,3,4) = \quad & [\Psi^{[J]}_{\beta}(1,2,3)\varphi^{[j]}(4)]^{[I]}_{M} \\ & - [\Psi^{[J]}_{\beta}(1,2,4)\varphi^{[j]}(3)]^{[I]}_{M} \\ & + [\Psi^{[J]}_{\beta}(1,3,4)\varphi^{[j]}(2)]^{[I]}_{M} \\ & - [\Psi^{[J]}_{\beta}(2,3,4)\varphi^{[j]}(1)]^{[I]}_{M},\end{aligned} \tag{6.26}$$

where from above

$$\Psi^{[J]}_{\beta M}(1,2,3) = \sum_{K} (j^2K;\ j|\}j^3\beta J)\ [\Psi^{[K]}(1,2)\varphi^{[j]}(3)]^{[I]}_{M}, \tag{6.27}$$

with K even.

In order to evaluate the norm overlaps

$$N^I_{\beta J\ \beta' J'} = \frac{1}{\hat{I}}\left[\Phi^{[I]}_{\beta J}\middle|\Phi^{[I]}_{\beta' J'}\right], \tag{6.28}$$

we proceed as before. The norm matrix elements are sums of 16 terms. We compute the four terms which correspond to particle index permutation in the ket. The 12 other terms are generated by particle permutation in the bra and yield the same contributions, hence a factor 4 in the final expression. The recouplings needed to bring back the particle indices of each term of the ket (6.26) into the order (1,2,3,4) is performed in the graph 6.7 which yields

$$\Phi^{[I]}_{\beta JM}(1,2,3,4) = \sum_{\beta' J'} [\Psi^{[J']}_{\beta'}(1,2,3)\varphi^{[j]}(4)]^{[I]}_M$$

$$\times \left(\delta_{JJ'}\delta_{\beta\beta'} - \sum_{KK'} (j^2K;\ j|\}j^3\beta J)(j^2J';j|\}j^3\beta' J') \right.$$

$$\left. \times \begin{bmatrix} K & j & J \\ j & 0 & j \\ J' & j & I \end{bmatrix} \left(\delta_{KK'} - (1+(-)^{K'}) \begin{bmatrix} j & j & K \\ j & 0 & j \\ K' & j & J' \end{bmatrix}\right) \right). \tag{6.29}$$

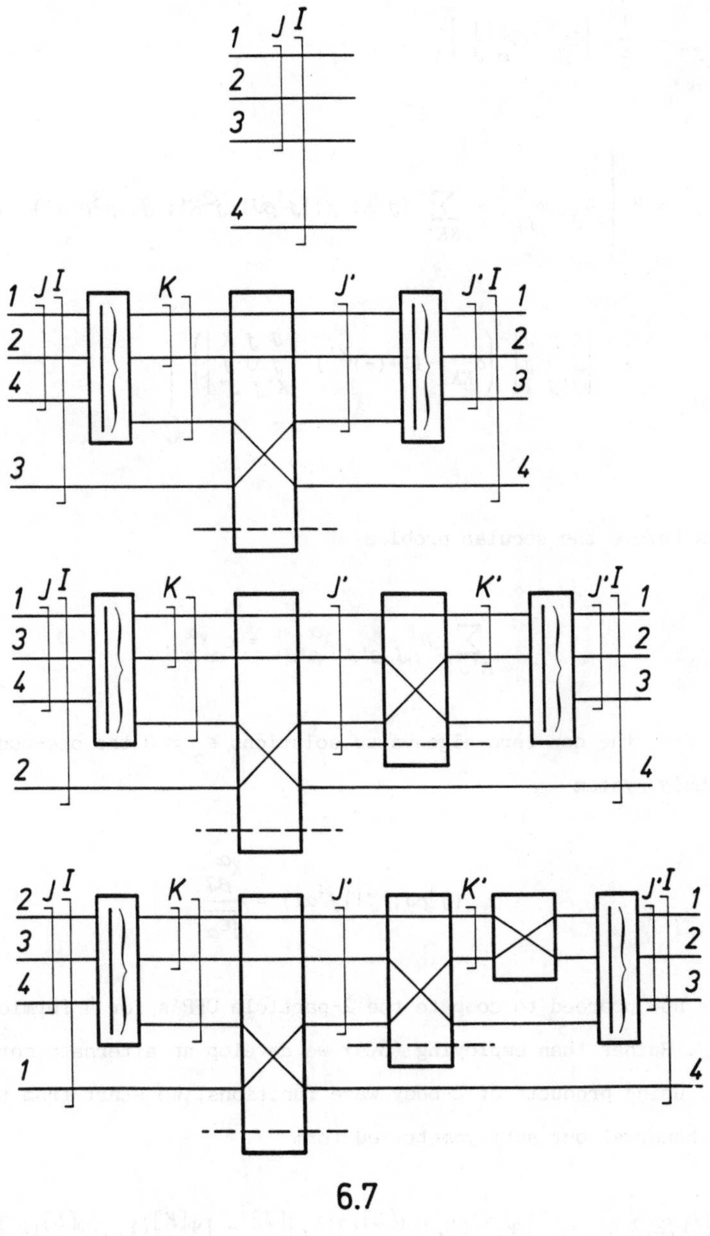

6.7

Now the detailed expression of the overlap matrix readily can be obtained

$$N^I_{\beta J\,\beta' J'} = \frac{1}{\hat{I}} \left[\Phi^{[I]}_{\beta J} | \Phi^{[I]}_{\beta' J'}\right]$$

$$= 4 \left(\delta_{JJ'} \delta_{\beta\beta'} - \sum_{KK'} (j^2K;\, j|\}j^3\beta J)(j^2K';\, j|\}j^3\beta' J') \right.$$

$$\left. \times \begin{bmatrix} K & j & J \\ j & 0 & j \\ J' & j & I \end{bmatrix} \left(\delta_{KK'} - (1+(-)^{K'}) \begin{bmatrix} J & j & K \\ j & 0 & j \\ K' & j & J' \end{bmatrix} \right) \right). \tag{6.30}$$

As before the secular problem

$$\sum_{\beta' J'} N^I_{\beta J\,\beta' J'} X^\alpha_{\beta' J'} = \epsilon_\alpha X^\alpha_{\beta J}, \tag{6.31}$$

yields from the non-zero eigenvalue solutions $\epsilon_\alpha \neq 0$ the one-body CFP's of the 4-body system

$$(j^3\beta J;\, j|\}j^4\alpha I) = \frac{X^\alpha_{\beta J}}{\sqrt{\epsilon_\alpha}}. \tag{6.32}$$

We now proceed to compute the 2-particle CFP's for 4 fermions in a shell j. Rather than employing (6.9) we develop an alternate more direct method, using products of 2-body wave functions. We start from the non-orthonormal but antisymmetrized form

$$\begin{aligned}\Phi^{[I]}_{KLM}(1,2,3,4) = \; & [\Psi^{[K]}(1,2)\Psi^{[L]}(3,4)]^{[I]}_M - [\Psi^{[K]}(1,3)\Psi^{[L]}(2,4)]^{[I]}_M \\ & - [\Psi^{[K]}(1,4)\Psi^{[L]}(2,3)]^{[I]}_M - [\Psi^{[K]}(2,3)\Psi^{[L]}(1,4)]^{[I]}_M \\ & - [\Psi^{[K]}(2,4)\Psi^{[L]}(1,3)]^{[I]}_M - [\Psi^{[K]}(3,4)\Psi^{[L]}(1,2)]^{[I]}_M,\end{aligned} \tag{6.33}$$

with K and L even. The overlap matrix is

$$N^I_{KL\,K'L'} = \frac{1}{\hat{I}}\left[\Phi^{[I]}_{KL}|\Phi^{[I]}_{K'L'}\right] = 4\left(\delta_{KK'}\delta_{LL'}+(-)^I\delta_{KL'}\delta_{K'L} - 4\begin{bmatrix} j & j & K' \\ j & j & L' \\ K & L & I\end{bmatrix}\right). \quad (6.34)$$

Again in terms of the non-zero eigenvalue solutions of the secular problem defined by this matrix we obtain the 2-particle CFP's for the 4-body system

$$(j^2K;\ j^2L|\}j^4\alpha I) = \frac{X^{\alpha}_{KL}}{\sqrt{\epsilon_\alpha}}. \quad (6.35)$$

These calculational procedures in fact are generally applicable for computing the CFP's for n-body systems when the n-1 and/or n-2 CFP's are known.

Chapter 7

FOCK SPACE

The graphical method applies directly to the Fock space once properly phased annihilation and creation operators have been defined, i.e. both operator types must transform as contrastandard tensors. We only have to introduce a new symbol in the graphs which keeps track of the non-commuting character of the operators. Again we will work exclusively with invariant quantities for both the Fock space state vectors and operators. The final results are expressed as the product of an invariant matrix element, to be computed with the previous techniques, and a vacuum expectation value of an invariant product of creation and annihilation operators. The graph method yields a compact way to implement the Wick theorem and evaluate these expectation values, as we show by means of a few examples.

Contrastandard creation and annihilation operators

In Fock space annihilation operators a_{jm} are defined relative to a vacuum state $|0\rangle$ by

$$a_{jm}|0\rangle = 0, \tag{7.1}$$

for all states jm.

The creation operator for a particle in the state jm is the hermitian conjugate operator

$$|jm\rangle = a^{\dagger}_{jm}|0\rangle. \qquad (7.2)$$

These operators are defined to fulfill the commutation (anticommutation) relations

$$a_{j'm'}a^{\dagger}_{jm} \mp a^{\dagger}_{jm}a_{j'm'} = \delta_{jj'}\delta_{mm'}, \qquad (7.3)$$

where - is for bosons and + for fermions.

The vacuum expectation values or contractions of a product of Fock operators follows

$$\langle 0|a_{j'm'}a^{\dagger}_{jm}|0\rangle = \delta_{jj'}\delta_{mm'}, \qquad (7.4)$$

$$\langle 0|a_{j'm'}a_{jm}|0\rangle = \langle 0|a^{\dagger}_{j'm'}a_{jm}|0\rangle = \langle 0|a^{\dagger}_{j'm'}a^{\dagger}_{jm}|0\rangle = 0. \qquad (7.5)$$

To deal with all Fock operators as operators fulfilling our phase convention, we begin by identifying

$$a_{jm} = a^{(j)}_{m}, \qquad (7.6)$$

namely the annihilation operators transform as standard tensors, eq.(3.5). This assignment agrees with the conventional use of these operators.

With this identification we define a canonical transformation to properly phased annihilation and creation operators. We introduce newly phased annihilation operators

$$a^{[j]}_{m} = (-)^{j-m}\, a_{j-m}, \qquad (7.7)$$

and creation operators

$$\tilde{a}^{[j]}_{m} = (-)^{2j}\, a^{\dagger}_{jm}, \tag{7.8}$$

which fulfill our phase convention, eq.(3.7),

$$a^{[j]\dagger}_{m} = (-)^{j+m}\, \tilde{a}^{[j]}_{-m}. \tag{7.9}$$

Both operator types, the annihilation operator $a^{[\ ]}$ and the creation operator $\tilde{a}^{[\ ]}$, transform as contrastandard tensors, eq.(3.3).

The commutation (anticommutation) relations of eq.(7.3), substituting these new properly phased operators, are preserved and have the following form

$$(-)^{j+m+2j'}\left(a^{[j]}_{-m}\tilde{a}^{[j']}_{m'} - \epsilon\, \tilde{a}^{[j']}_{m'}a^{[j]}_{-m}\right) = \delta_{jj'}\delta_{mm'}, \tag{7.10}$$

or in coupled form, obtained by multiplication with the appropriate vector coupling coefficients and summing over the magnetic quantum numbers

$$[a^{[j]}\tilde{a}^{[j']}]^{[I]}_{M} + \epsilon(-)^{j+j'-I}[\tilde{a}^{[j']}a^{[j]}]^{[I]}_{M} = \epsilon\, \hat{j}\, \delta_{jj'}\delta_{I0}. \tag{7.11}$$

The phase factor ϵ is

$$\begin{aligned} \epsilon &= -1 \quad \text{for fermions,} \\ \epsilon &= +1 \quad \text{for bosons,} \end{aligned} \tag{7.12}$$

or here, where necessarily $j = j'$,

$$\epsilon = (-)^{2j} \quad \text{for both cases.} \tag{7.13}$$

The vacuum expectation values of eqs.(7.4) and (7.5) write in coupled form after replacement by the new operators (7.7) and (7.8)

$$\langle 0|[a^{[j']}\,\tilde{a}^{[j]}]^{[J]}|0\rangle = \epsilon\,\hat{j}\,\delta_{jj'}\delta_{J0}, \tag{7.14}$$

$$\langle 0|[\tilde{a}^{[j']}a^{[j]}]^{[J]}|0\rangle = \langle 0|[a^{[j']}a^{[j]}]^{[J]}|0\rangle = \langle 0|[\tilde{a}^{[j']}\tilde{a}^{[j]}]^{[J]}|0\rangle = 0. \tag{7.15}$$

From now on we will operate exclusively with these new properly phased operators, eqs.(7.7), (7.8), and use the contraction table (7.14)-(7.15).

Fock space graphs

We now may introduce a graphical representation for Fock space expressions. The new creation and annihilation operators $\tilde{a}^{[j]}$, $a^{[j]}$ having been defined to transform as contrastandard tensors, all previous graphical symbols and rules apply. Thus these operators are represented by horizontal lines with brackets indicating the appropriate couplings. Lines associated with creation operators $\tilde{a}^{[j]}$ are indicated by a ~ symbol. We only have to account for the commutation phase ϵ, eq.(7.12). This phase is represented on the graph by a circle around the crossings of the operator lines in the recoupling boxes. Thus the crossing box of fig.7.1 has the value v

$$v = \epsilon(-)^{j+j'-J}, \tag{7.16}$$

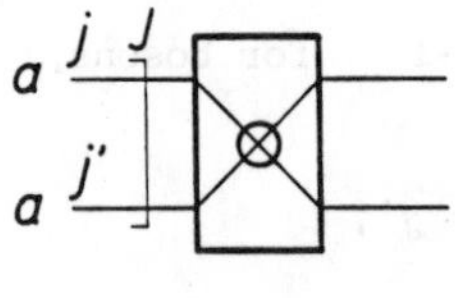

7.1

and the recoupling box of fig.7.2 the value

$$v = \epsilon \begin{bmatrix} j & k & J \\ \ell & m & K \\ P & Q & I \end{bmatrix}. \tag{7.17}$$

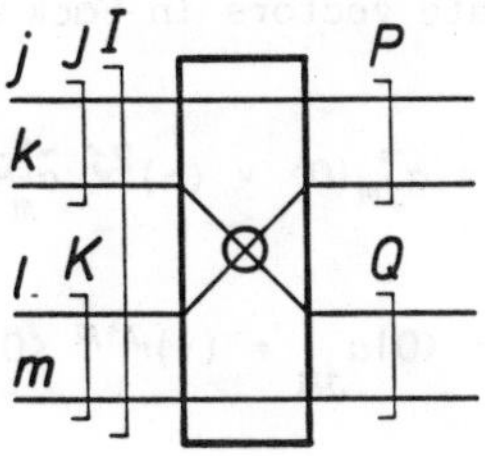

7.2

The invariant matrix element of the non-zero contraction, as given by eq.(7.14), is represented by the end-box of fig.7.3. Its value is that of the usual end-box but for the added factor ϵ. This factor is represented on the box by a circle. Hence the contraction symbol of fig.7.3 has the value

$$v = \epsilon \, \hat{j}. \tag{7.18}$$

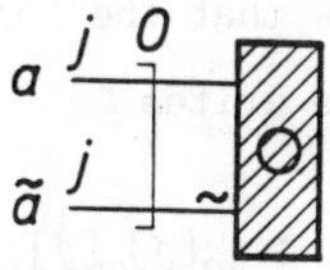

7.3

It is important to note that in this case the tilda line of the creation operator must enter the box at the bottom, in conformity with the order of the operators in the contraction, eq. (7.14).

As an example, the graphical representation of the commutation relations of eq.(7.11) is given by fig.7.4.

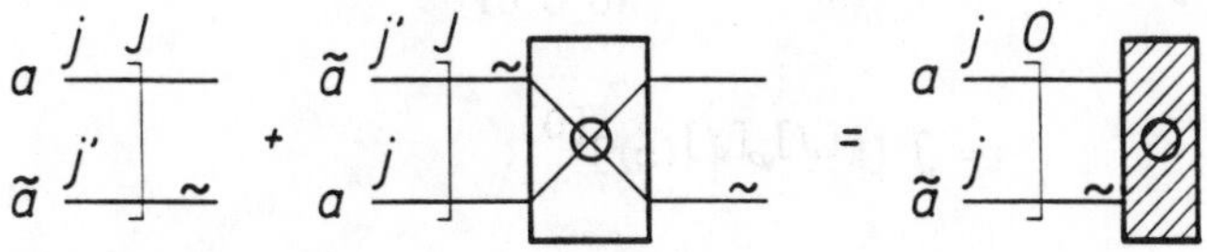

7.4

The single-particle state vectors in Fock space are

$$|jm\rangle = a^{\dagger}_{jm}|0\rangle = (-)^{2j}\, \tilde{a}^{[j]}_{m}|0\rangle \tag{7.19}$$

$$\langle jm| = \langle 0|a_{jm} = (-)^{j+m}\, \langle 0|a^{[j]}_{-m} \tag{7.20}$$

According to the invariant graph method we work with their invariant forms obtained as usual by introducing the state amplitudes $W^{[j]}_{m}$ defined in Chapter 2. Thus the state vectors write respectively,

$$|j\rangle = (-)^{2j}\, \hat{j}\, [W^{[j]}\tilde{a}^{[j]}]^{[0]}\, |0\rangle, \tag{7.21}$$

and

$$\langle j| = \hat{j}\, \langle 0|\, [\tilde{W}^{[j]}a^{[j]}]^{[0]}. \tag{7.22}$$

As an exercise let us show that the Fock space operator for creating a wave function in ordinary space writes

$$\varphi_{j}(x) = \hat{j}\, [a^{[j]}\varphi^{[j]}(x)]^{[0]}. \tag{7.23}$$

Its matrix element between the ket $|jm\rangle$ and the vacuum $\langle 0|$

$$\langle 0|\varphi_{j}(x)|jm\rangle = (-)^{2j}\, \hat{j}^{2}\langle 0|\left[a^{[j]}\varphi^{[j]}(x)\right]^{[0]}\left[W^{[j]}\tilde{a}^{[j]}\right]^{[0]}|0\rangle, \tag{7.24}$$

with the value $\left|W^{[j]}_{m'}\right| = \delta_{mm'}$, yields from the graph 7.5

$$\langle 0|\varphi_{j}(x)|jm\rangle = (-)^{2j}\, \hat{j}^{2}(-)^{2j}\begin{bmatrix} j & j & 0\\ j & j & 0\\ 0 & 0 & 0\end{bmatrix}(-)^{2j}\, \epsilon\, \hat{j}\, [W^{[j]}\varphi^{[j]}(x)]^{[0]}$$

$$= \hat{j}\, [W^{[j]}\varphi^{[j]}(x)]^{[0]}. \tag{7.25}$$

This is the usual invariant form for the wave function $\varphi_m^{[j]}(x)$, eq.(2.10).

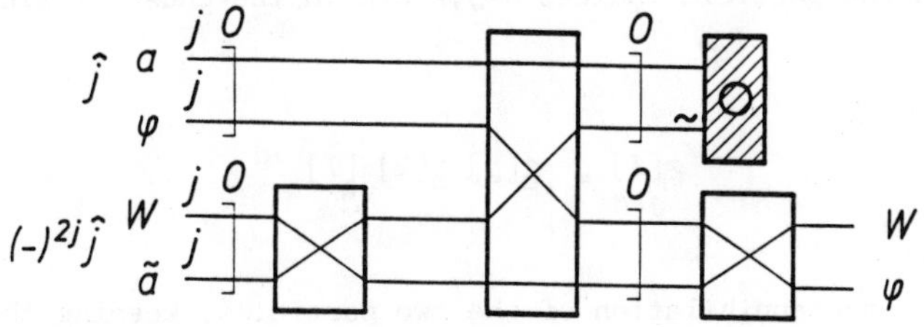

7.5

We now compute the norm of the state $|j\rangle$. From eqs.(7.21) and (7.22)

$$\langle j|j\rangle = (-)^{2j}\,\hat{j}^2\langle 0|\,[\tilde{W}^{[j]}a^{[j]}]^{[0]}\,[W^{[j]}\tilde{a}^{[j]}]^{[0]}\,|0\rangle. \tag{7.26}$$

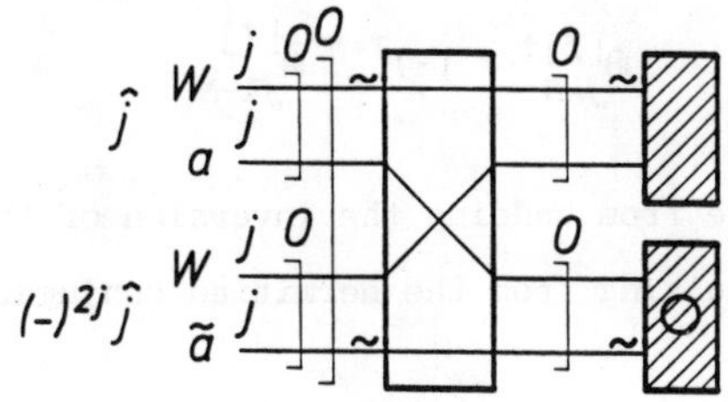

7.6

It is read off the graph 7.6, which yields

$$\langle j|j\rangle = (-)^{2j}\,\hat{j}^2 \begin{bmatrix} j & j & 0 \\ j & j & 0 \\ 0 & 0 & 0 \end{bmatrix} \frac{1}{\hat{j}}\,\epsilon\,\hat{j} = 1. \tag{7.27}$$

Two-particle states

The bras and kets for many-particles states in Fock space are products of creation and annihilation operators always written in a lexicographic order $j,k,\ell\ldots$ in order to fix the permutation phases.

As the simplest example we consider the case of two particles in non-identical shells. We introduce the coupled creation operator $\tilde{\mathbb{A}}^{[I]}_{n\alpha M}$, where n=2 and the particle indices α=j,k are in the chosen lexicographic order

$$\tilde{\mathbb{A}}^{[I]}_{jkM} = [\tilde{a}^{[j]}\ \tilde{a}^{[k]}]^{[I]}_{M}. \tag{7.28}$$

Then, for the annihilation of the two particles, keeping the lexicographic order j,k in this definition

$$\mathbb{A}^{[I]}_{jkM} = \epsilon\ [a^{[j]} a^{[k]}]^{[I]}_{M}. \tag{7.29}$$

The phase ϵ defined in eq.(7.12), i.e. ϵ = +1 for bosons and ϵ = -1 for fermions, insures that

$$\mathbb{A}^{[I]\dagger}_{jkM} = (-)^{I+M}\ \tilde{\mathbb{A}}^{[I]}_{jk-M}. \tag{7.30}$$

The phase ϵ arises here from undoing the inversion of the order of the particle operators resulting from the hermitian conjugation.

We may also verify that the vacuum expectation value of the coupled operators for two particles is from fig.7.7

$$\langle 0|\ [\mathbb{A}^{[I]}_{jk}\ \ \tilde{\mathbb{A}}^{[I]}_{jk}]^{[0]}\ |0\rangle = \epsilon^{4}\begin{bmatrix} j & k & I \\ j & k & I \\ 0 & 0 & 0 \end{bmatrix}\ \hat{j}\ \hat{k} = \hat{I}. \tag{7.31}$$

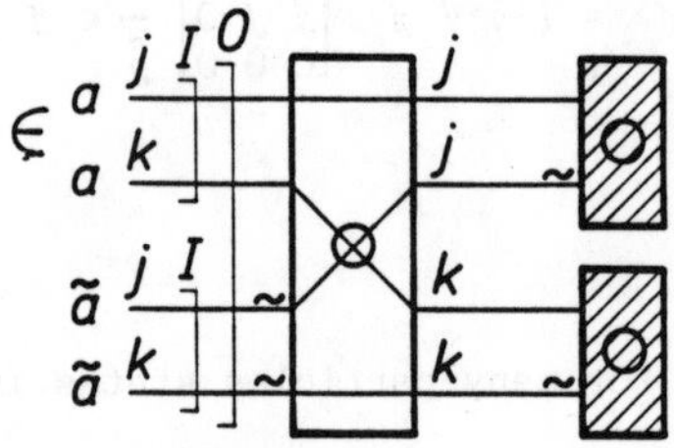

7.7

The corresponding normalized two particle states in invariant form are now given as

$$|jk\ ;\ I\rangle = (-)^{2I}\ \hat{I}\ \left[W^{[I]}\ \tilde{\mathbb{A}}^{[I]}_{jk}\right]^{[0]}\ |0\rangle, \tag{7.32}$$

$$\langle jk\ ;\ I| = \langle 0|\ \hat{I}\ \left[\tilde{W}^{[I]}\mathbb{A}^{[I]}_{jk}\right]^{[0]}. \tag{7.33}$$

The factor $(-)^{2I}$ generalizes the definition of eq.(7.21). Here for n=2 its value is of course +1 (I integer). However we keep it explicitly in front of the ket since this phase factor is required for proper normalization in the case of an odd number of fermions.

When the two particles are in the same shell j, the operators $\tilde{\mathbb{A}}^{[I]}_{jjM}$ yielding normalized states are of the form

$$\tilde{\mathbb{A}}^{[I]}_{jjM} = \frac{1}{\sqrt{2}}\ [\tilde{a}^{[j]}\tilde{a}^{[j]}]^{[I]}_{M}, \tag{7.34}$$

with I even.

Fock space two- and one-body operators

Operators in Fock space always will be written as a sum of products of two invariants : an invariant matrix element and an invariant constructed from creation and annihilation operators. In order to illustrate this we will develop in detail their construction for a 2-body scalar operator.

We start from the form of a two-body scalar operator V in the uncoupled m-scheme. In terms of the ordinary creation and annihilation operators, eqs.(7.1) and (7.2), its Fock expression is

$$V = \frac{1}{2}\sum_{\substack{ijkl\\ m's}}\int \varphi^{\dagger}_{im_i}(1)\varphi^{\dagger}_{jm_j}(2)\ \mathcal{V}(1,2)\ \varphi_{km_k}(1)\varphi_{\ell m_\ell}(2)\ :a^{\dagger}_{im_i}a^{\dagger}_{jm_j}a_{\ell m_\ell}a_{km_k}:\ , \tag{7.35}$$

where the double dots indicate the normal product of the operators. We note the order of the annihilation operators in the normal product required for fermions and the factor $\frac{1}{2}$ which corrects for overcounting in view of the unrestricted summations.

Using the phase convention (3.7) and substituting the properly phased operators, eqs.(7.7) and (7.8), we obtain

$$V = \epsilon\ \frac{1}{2} \sum_{\substack{ijkl\\ m's}} \int (-)^{i+m_i} \tilde{\varphi}^{[i]}_{-m_i} (-)^{j+m_j} \tilde{\varphi}^{[j]}_{-m_j}\ V\ \varphi^{[k]}_{m_k}\ \varphi^{[\ell]}_{m_\ell}$$

$$\times :(-)^{2i} \tilde{a}^{[i]}_{m_i} (-)^{2j} \tilde{a}^{[j]}_{m_j} (-)^{k+m_k} a^{[k]}_{-m_k} (-)^{\ell+m_\ell} a^{[\ell]}_{-m_\ell}: . \qquad (7.36)$$

The phase ϵ, as defined in eq.(7.12), arises from reordering the annihilation operators so as to have the same order of the indices in the normal product and in the matrix element. This ordering will prove to be convenient. We recognize the following product of invariants after performing the sum over the m's

$$V = \epsilon\ \frac{1}{2} \sum_{ijk\ell} \int :\hat{i}[\tilde{\varphi}^{[i]} \tilde{a}^{[i]}]^{[0]} \hat{j}[\tilde{\varphi}^{[j]} \tilde{a}^{[j]}]^{[0]} V^{[0]}\ \hat{k}[\varphi^{[k]} a^{[k]}]^{[0]}\ \hat{\ell}[\varphi^{[\ell]} a^{[\ell]}]^{[0]}: .$$

$$(7.37)$$

Thus the operator V can be written as a sum of products of two invariants

$$V = \frac{1}{2} \epsilon \sum_{ijk\ell} \sum_{I} \left[[\varphi^{[i]}(1) \varphi^{[j]}(2)]^{[I]} \middle| V^{[0]}(1,2) \middle| [\varphi^{[k]}(1) \varphi^{[\ell]}(2)]^{[I]} \right]$$

$$\times :\ [[\tilde{a}^{[i]} \tilde{a}^{[j]}]^{[I]} [a^{[k]} a^{[\ell]}]^{[I]}]^{[0]} : ,$$

$$(7.38)$$

as shown by the recoupling graph of fig.7.8 for the integrand of eq.(7.37) which yields

$$\hat{\imath}\ \hat{\jmath}\ \hat{k}\ \hat{\ell} \begin{bmatrix} i & i & 0 \\ j & j & 0 \\ I & I & 0 \end{bmatrix} \begin{bmatrix} k & k & 0 \\ \ell & \ell & 0 \\ I & I & 0 \end{bmatrix} \begin{bmatrix} I & I & 0 \\ I & I & 0 \\ 0 & 0 & 0 \end{bmatrix} = 1. \tag{7.39}$$

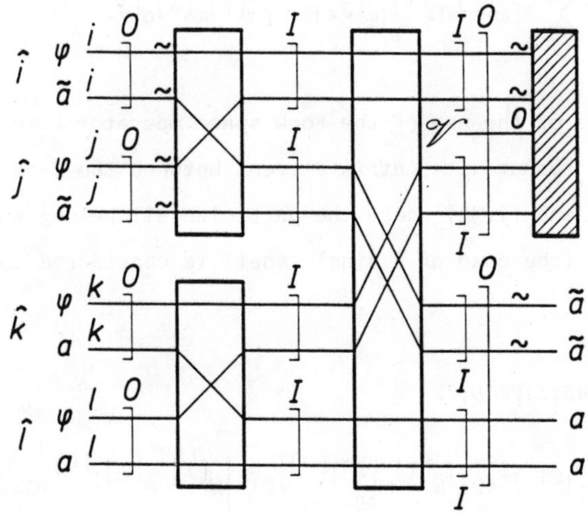

7.8

We note again that in the coupled invariant form of eq.(7.38) the order of the state indices i,j,k,ℓ has been chosen to be the same both in the invariant matrix element and in the invariant normal product of Fock operators, hence the phase ϵ. The invariant matrix elements of $\mathcal{V}^{[0]}$ in eq.(7.38) are known : they are evaluated with the method of the previous Chapters. The invariant normal product will enter the vacuum expectation values of Fock operators as will be discussed below.

The Fock space expressions for multipole operators

$$\Omega_\lambda = \hat{\lambda}\ [\omega^{[\lambda]}\Omega^{[\lambda]}]^{[0]} \tag{7.40}$$

likewise can be written as products of invariants. By following the steps leading to eq.(7.38) we obtain for a one-body multipole operator

$$\Omega_\lambda = \sum_{ij}\ [\varphi^{[i]}|\Omega^{[\lambda]}|\varphi^{[j]}]\ :\ [\tilde{a}^{[i]}\omega^{[\lambda]}a^{[j]}]^{[0]}:\ . \tag{7.41}$$

As an example of the use of the Fock space operator V in the form of eq.(7.38), we now compute its matrix element between the 2-particle bra and ket of eqs.(7.32) and (7.33) where the particles all belong to different shells A, B, C, D (the case of a single shell is considered in the next Section),

$$M = \langle AB;I|V|CD;I\rangle$$

$$= (-)^{2I}\ \hat{I}^2\langle 0|\left[\tilde{W}^{[I]}\mathbb{A}^{[I]}_{AB}\right]^{[0]}|V|\left[W^{[I]}\ \tilde{\mathbb{A}}^{[I]}_{CD}\right]^{[0]}|0\rangle. \tag{7.42}$$

The calculation involves the evaluation of the vacuum expectation value

$$\langle 0|G|0\rangle = \epsilon(-)^{2I}\langle 0|\ \left[\tilde{W}^{[I]}[a^{[A]}a^{[B]}]^{[I]}\right]^{[0]}$$

$$\times\ :\ \left[[\tilde{a}^{[i]}\ \tilde{a}^{[j]}]^{[I']}[a^{[k]}a^{[\ell]}]^{[I']}\right]^{[0]}:$$

$$\times\ \left[W^{[I]}[\tilde{a}^{[C]}\ \tilde{a}^{[D]}]^{[I]}\right]^{[0]}\ |0\rangle. \tag{7.43}$$

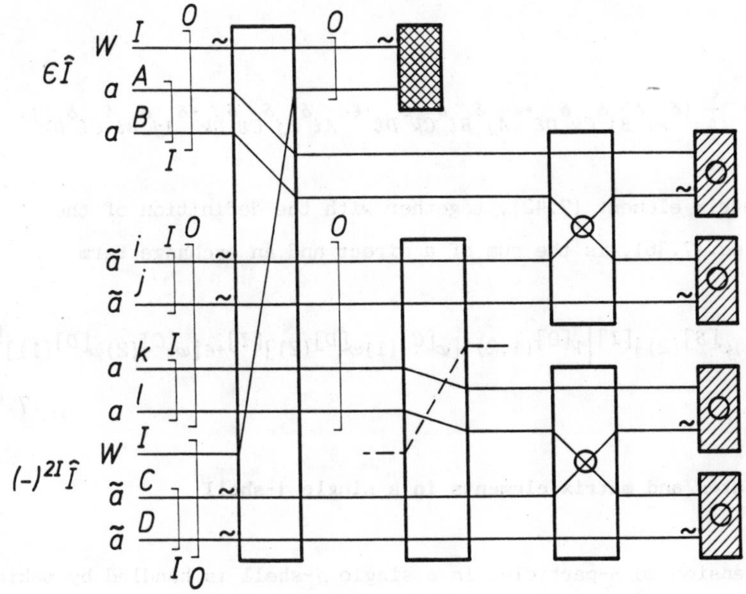

7.9

The bra and ket indices A,B and C,D are each in lexicographic order. In contrast, the operator indices i,j,k,ℓ in the normal product run indepentently of each other, eq.(7.35).Thus there are four distinct systems of contraction, one of which is drawn on fig.7.9, introducing the circles at the crossing points of the operator lines in the recoupling boxes, the circles in the vacuum expectation end-boxes and the selection rule $I' = I$. This graph yields the contribution

$$g = (-)^{2I}\epsilon^7\ \hat{I}^2 \begin{bmatrix} I & I & 0 \\ I & I & 0 \\ 0 & 0 & 0 \end{bmatrix} \frac{1}{\hat{I}} \begin{bmatrix} I & I & 0 \\ 0 & I & I \\ I & 0 & I \end{bmatrix} \begin{bmatrix} A & B & I \\ i & j & I \\ 0 & 0 & 0 \end{bmatrix}$$

$$\times \begin{bmatrix} k & \ell & I \\ C & D & I \\ 0 & 0 & 0 \end{bmatrix} \hat{A}\ \delta_{Ai}\ \hat{B}\ \delta_{Bj}\ \hat{C}\ \delta_{Ck}\ \hat{D}\ \delta_{D\ell}$$

$$= \frac{\epsilon}{\hat{I}}\ \delta_{Ai}\delta_{Bj}\delta_{Ck}\delta_{D\ell}. \qquad (7.44)$$

Together with the three other contraction graphs, which are similar except for simple permutations of the lines, we get for the complete expectation value (7.43)

$$\langle 0|G|0\rangle = \frac{\epsilon}{\hat{I}}(\delta_{Ai}\delta_{Bj}\delta_{Ck}\delta_{D\ell}+\epsilon\,\delta_{Aj}\delta_{Bi}\delta_{Ck}\delta_{D\ell}+\epsilon\,\delta_{Ai}\delta_{Bj}\delta_{C\ell}\delta_{Dk}+\delta_{Aj}\delta_{Bi}\delta_{C\ell}\delta_{Dk}).$$

Hence the matrix element (7.42), together with the definition of the operator V, eq.(7.36), is the sum of a direct and an exchange term

$$M = \left[[\varphi^{[A]}(1)\varphi^{[B]}(2)]^{[I]}\middle|v^{[0]}(1,2)\middle|[\varphi^{[C]}(1)\varphi^{[D]}(2)]^{[I]}+\epsilon[\varphi^{[C]}(2)\varphi^{[D]}(1)]^{[I]}\right]. \tag{7.45}$$

Many-body states and matrix elements in a single j-shell

The extension to n-particles in a single j-shell is handled by making use of the fractional parentage coefficients introduced in Chapter 6.

The n-particle Fock creation operator $\tilde{A}^{[I]}_{n\alpha M}$ for the coupled state j^n of total angular momentum I, M and quantum numbers α is defined recursively by expansions with CFP's. Separating out one particle we have in terms of one-particle CFP's, eq.(6.5),

$$\tilde{A}^{[I]}_{n\alpha M} = \frac{1}{\sqrt{n}}\sum_{\beta K}(j^{n-1}\beta K;\ j|\}j^n\alpha I)\ [\tilde{A}^{[K]}_{n-1\beta}\ \tilde{a}^{[j]}]^{[I]}_M, \tag{7.46}$$

or with 2-particle CFP's, eq.(6.6),

$$\tilde{A}^{[I]}_{n\alpha M} = \sqrt{\frac{2}{n(n-1)}}\sum_{\beta KL}(j^{n-2}\beta K;\ j^2L||\}j^n\alpha I)\ [\tilde{A}^{[K]}_{n-2\beta}\ \tilde{A}^{[L]}_{2}]^{[I]}_M. \tag{7.47}$$

Note the multiplicity factors. They arise from the fact that the operator products $\tilde{A}^{[K]}_{n-1\beta}\ \tilde{a}^{[j]}$ and $\tilde{A}^{[L]}_{n-2\beta}\ \tilde{A}^{[K]}_{2}$ generate symmetrized (antisymmetrized) states which are not normalized, while the definitions of the CFP's, eq.(6.5) and eq.(6.6), were given in terms of normalized states.

The corresponding expressions for the coupled annihilation operators are obtained from the hermitian conjugation relations. Thus we obtain from (7.46)

$$\mathbb{A}^{[I]}_{n\alpha M} = \frac{\epsilon^{n-1}}{\sqrt{n}} \sum_{\beta K} (j^{n-1}\beta K;\ j|\}j^n\alpha I)\ [\mathbb{A}^{[K]}_{n-1\beta}\ a^{[j]}]^{[I]}_{M}. \tag{7.48}$$

The reordering of the last operator to the right contributes the phase factor ϵ^{n-1}. Similarly we have from (7.47)

$$\mathbb{A}^{[I]}_{n\alpha M} = \sqrt{\frac{2}{n(n-1)}} \sum_{\beta KL} (j^{n-2}\beta K;\ j^2 L||\}j^n\alpha I)\ [\mathbb{A}^{[K]}_{n-2\beta}\ \mathbb{A}^{[L]}_{2}]^{[I]}_{M}. \tag{7.49}$$

The vacuum expectation value of the n-particle Fock operators, taking into account the normalization (6.7) or (6.8) of the CFP's, is

$$\langle 0|\ \left[\mathbb{A}^{[I]}_{n\alpha}\ \tilde{\mathbb{A}}^{[I]}_{n\alpha}\right]^{[0]}\ |0\rangle = \epsilon^n\ \hat{I} = (-)^{2I}\ \ \hat{I}. \tag{7.50}$$

Here the remaining phase factor arises from the n contraction end-boxes, and of course $\epsilon^n = (-)^{2I}$. Thus the general definition of the value of an overlap end-box, eq.(7.14), is maintained.

The invariant forms of the coupled many-particle Fock bras and kets now are given in the usual way by

$$\begin{aligned} |n\alpha;I\rangle &= (-)^{2I}\ \hat{I}\ [W^{[I]}\tilde{\mathbb{A}}^{[I]}_{n\alpha}]^{[0]}\ |0\rangle, \\ \langle n\alpha;I| &= \hat{I}\ \langle 0|\ [\tilde{W}^{[I]}\mathbb{A}^{[I]}_{n\alpha}]^{[0]}. \end{aligned} \tag{7.51}$$

Note that in these Fock space invariant forms the phase $(-)^{2I}$ again is associated with the ket. We let the reader verify by drawing the appropriate graph that, of course, there holds the norm

$$\langle n\alpha;I|n\alpha;I\rangle = 1. \tag{7.52}$$

A many-body multipole operator Ω_λ acting in a single j-shell will be of the general form,

$$\Omega_\lambda = \sum_{\substack{KL\\ \gamma\delta}} \left[F_\gamma^{[K]}\middle|\Omega^{[\lambda]}\middle|F_\delta^{[L]}\right] : \left[[\tilde{a}\ldots\tilde{a}]_\gamma^{[K]}\ \omega^{[\lambda]}\ [a\ldots a]_\delta^{[L]}\right]^{[0]} : . \qquad (7.53)$$

The number of annihilation and of creation operators in (7.53) of course does not have to be the same.

The matrix element of the operator (7.53) between many-particle states (7.51) writes

$$M = \langle n\alpha J|\Omega_\lambda|n'\beta I\rangle$$

$$= (-)^{2I}\ \hat{I}\ \hat{J} \sum_{\substack{KL\\ \gamma\delta}} \left[F_\gamma^{[K]}\middle|\Omega^{[\lambda]}\middle|F_\delta^{[L]}\right]$$

$$\times \langle 0|\ [\tilde{W}_\alpha^{[J]}\mathbb{A}_{n\alpha}^{[J]}]^{[0]} : \left[[\tilde{a}\ldots\tilde{a}]_\gamma^{[K]}\ \omega^{[\lambda]}\ [a\ldots a]_\delta^{[L]}\right]^{[0]} : [W_\beta^{[I]}\tilde{\mathbb{A}}_{n'\beta}^{[I]}]^{[0]}\ |0\rangle, \qquad (7.54)$$

and is evaluated by drawing the graph 7.10 which yields

$$M = \frac{(-)^{2I}}{\hat{\lambda}}\ (-)^{K+\lambda-L}\left[F_\gamma^{[K]}|\Omega^{[\lambda]}|F_\delta^{[L]}\right]\ [\tilde{W}_\alpha^{[I]}\omega^{[\lambda]}W_\beta^{[I]}]^{[0]}\ \langle 0|G|0\rangle. \qquad (7.55)$$

The Fock space aspects are thus contained in the evaluation of the vacuum expectation value

$$\langle 0|G|0\rangle = \langle 0|\left[\mathbb{A}_{n\alpha}^{[J]} : \left[[\tilde{a}\ldots\tilde{a}]_\gamma^{[K]}[a\ldots a]_\delta^{[L]}\right]^{[\lambda]} : \tilde{\mathbb{A}}_{n'\beta}^{[I]}\right]^{[0]} |0\rangle. \qquad (7.56)$$

Some frequently encountered cases are treated below.

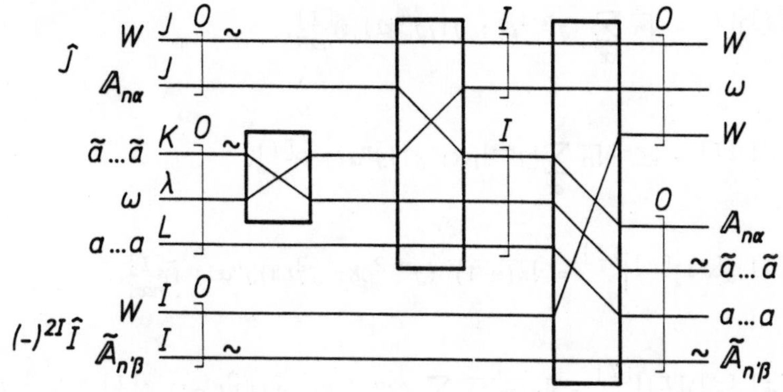

7.10

As discussed in Chapter 4, the amplitude factor $[\underset{\alpha}{\tilde{W}}^{[J]}\omega^{[\lambda]}\underset{\beta}{W}^{[I]}]^{[0]}$ in eq.(7.55) depends on the polarizations $\underset{\alpha}{W}^{[J]}$, $\underset{\beta}{W}^{[I]}$ of the states and on the amplitudes $\omega^{[\lambda]}$ of the operator. In the case of a scalar operator, $\omega^{[0]} = 1$, this factor takes the well-known simple value,

$$[\,\tilde{W}^{[J]}\omega^{[0]}W^{[I]}]^{[0]} = \frac{1}{\hat{I}}\,\delta_{IJ}. \tag{7.57}$$

Some more general examples of amplitude invariant products are given in Chapter 10.

Evaluation of vacuum expectation values

From the above, the calculation of many-body matrix elements in Fock space requires the evaluation of the general vacuum expectation value (7.56). We shall now evaluate it for the most often encountered cases of one-body and two-body operators.

Several relations are useful ; they are obtained from inverting eqs.(7.46) through (7.49) by means of the CFP's orthogonality relations (6.7) and (6.8), which here is possible since the Fock space operators ensure the proper symmetry of each individual term in the sums. We get

$$[\tilde{\mathbb{A}}^{[K]}_{n-1\beta}\ \tilde{a}^{[j]}]^{[I]}_{M} = \sqrt{n} \sum_{\alpha} (j^{n-1}\beta K;\ j|\}j^{n}\alpha I)\ \tilde{\mathbb{A}}^{[I]}_{n\alpha M}, \tag{7.58}$$

$$[\mathbb{A}^{[K]}_{n-1\beta}\ a^{[j]}]^{[I]}_{M} = \epsilon^{n-1}\sqrt{n} \sum_{\alpha}(j^{n-1}\beta K;\ j|\}j^{n}\alpha I)\ \mathbb{A}^{[I]}_{n\alpha M}, \tag{7.59}$$

$$\left[\tilde{\mathbb{A}}^{[K]}_{n-2\beta}[\tilde{a}^{[j]}\ \tilde{a}^{[j]}]^{[L]}\right]^{[I]}_{M} = \sqrt{n(n-1)}\ (j^{n-2}\beta K;\ j^{2}L||\}j^{n}\alpha I)\ \tilde{\mathbb{A}}^{[I]}_{n\alpha M}, \tag{7.60}$$

$$\left[\mathbb{A}^{[K]}_{n-2\beta}[a^{[j]}a^{[j]}]^{[L]}\right]^{[I]}_{M} = \sqrt{n(n-1)} \sum_{\alpha} (j^{n-2}\beta K;\ j^{2}L||\}j^{n}\alpha I)\ \mathbb{A}^{[I]}_{n\alpha M}. \tag{7.61}$$

In these expressions, the difference between the coupled operators $[a^{[j]}a^{[j]}]^{[L]}$, which are not normalized, and the operators $\mathbb{A}^{[L]}_{jj}$, which are eq.(7.34), contributes a factor $\sqrt{2}$.

In the evaluation of vacuum expectation values, as a general rule, one always has to use the eqs.(7.58) - (7.61) to first collect separately all creation operators and all annihilation operators into the properly symmetrized (antisymetrised) operators $\tilde{\mathbb{A}}$, $\mathbb{A}$ respectively, before performing the vacuum expectation overlaps. Only in this way the proper symmetry of the operator products - and the correct value of the overall vacuum expectation value - is achieved.

Now we evaluate some specific examples. For a unit operator we get

$$\langle 0|G_{0}|0\rangle = \langle 0| \left[\mathbb{A}^{[I]}_{n\alpha}\ \tilde{\mathbb{A}}^{[I]}_{n\alpha}\right]^{[0]} |0\rangle = (-)^{2I}\ \hat{I}. \tag{7.62}$$

For a one-particle creation operator,

$$\langle 0|G_{+1}|0\rangle = \langle 0| \left[\mathbb{A}^{[J]}_{n\alpha}\ :\tilde{a}^{[j]}:\ \tilde{\mathbb{A}}^{[I]}_{n-1\beta}\ \right]^{[0]} |0\rangle, \tag{7.63}$$

the application of eq.(7.58) and of the normalization relation (7.31) of the operators $\mathbb{A}^{[J]}_{n\alpha}$ requires a recoupling calculated with the graph 7.11

which reads (the double circle in the crossing box represents the crossing of one particle line with the n-1 others)

$$\langle 0|G_{+1}|0\rangle = \sqrt{n}\,\epsilon^{n-1}(-)^{j+I-J}\,(j^{n-1}\beta I;\ j|\}j^{n}\alpha J)(-)^{2J}\,\hat{J}. \tag{7.64}$$

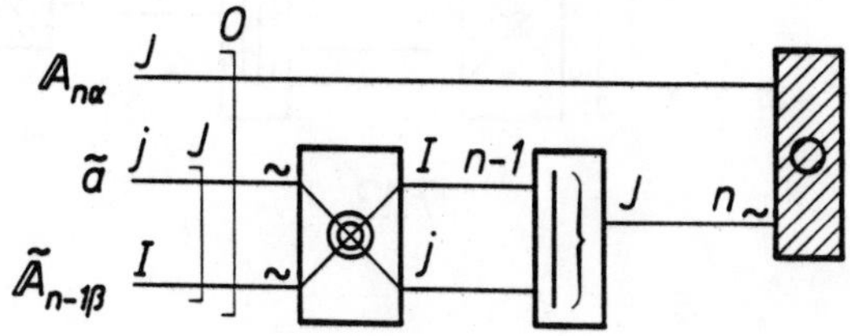

7.11

For a one-particle annihilation operator

$$\langle 0|G_{-1}|0\rangle = \langle 0|\left[\mathbb{A}^{[J]}_{n-1\alpha}:a^{[j]}:\tilde{\mathbb{A}}^{[I]}_{n\beta}\right]^{[0]}|0\rangle, \tag{7.65}$$

using eq.(7.59) the graph 7.12 yields

$$\langle 0|G_{-1}|0\rangle = \sqrt{n}\,\epsilon^{n-1}\,(j^{n-1}\alpha J;\ j|\}j^{n}\beta I)(-)^{2I}\,\hat{I}. \tag{7.66}$$

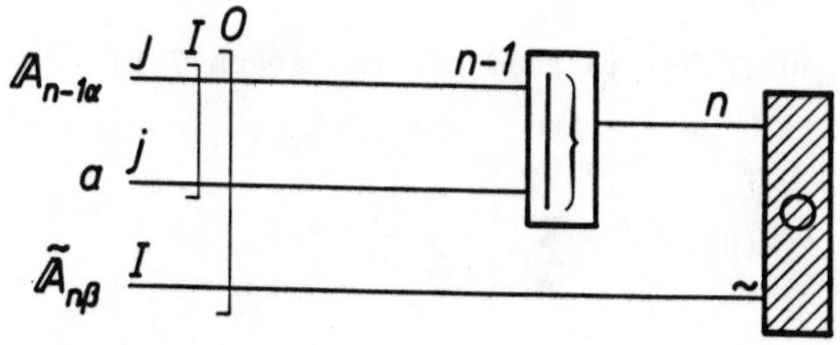

7.12

For the pair creation operator

$$\langle 0|G_{+2}|0\rangle = \langle 0|\left[\mathbb{A}^{[J]}_{n\alpha}:[\tilde{a}^{[j]}\,\tilde{a}^{[j]}]^{[L]}:\tilde{\mathbb{A}}^{[I]}_{n-2\beta}\right]^{[0]}|0\rangle, \tag{7.67}$$

we get from eq.(6.60) and fig.7.13

$$\langle 0|G_{+2}|0\rangle = \sqrt{n(n-1)}\,(-)^{L+I-J}(j^{n-2}\beta I;j^2L)|\}j^n\alpha J)(-)^{2J}\,\hat{J}. \qquad (7.68)$$

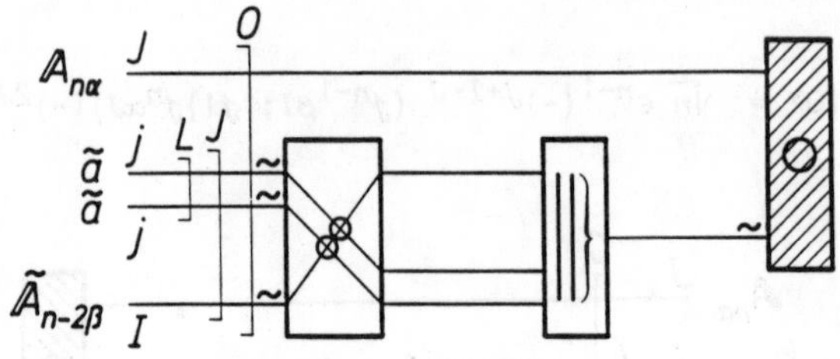

7.13

For the pair annihilation operator

$$\langle 0|G_{-2}|0\rangle = \langle 0|\left[\mathbb{A}^{[J]}_{n-2\alpha} : [a^{[j]}a^{[j]}]^{[L]} : \tilde{\mathbb{A}}^{[I]}_{n\beta}\right]^{[0]}|0\rangle, \qquad (7.69)$$

we get similarly, eq.(7.61),

$$\langle 0|G_{-2}|0\rangle = \sqrt{n(n-1)}\;(j^{n-2}\alpha J;\; j^2L\|\}j^n\beta I)\;(-)^{2I}\;\hat{I}. \qquad (7.70)$$

Finally the mean value of the mixed pair operator

$$\langle 0|G^1_0|0\rangle = \langle 0|\left[\mathbb{A}^{[J]}_{n\alpha} : [\tilde{a}^{[j]}a^{[j]}]^{[L]} : \tilde{\mathbb{A}}^{[I]}_{n\beta}\right]^{[0]}|0\rangle, \qquad (7.71)$$

using eqs.(7.58) and (7.59) is read off the graph 7.14

$$\langle 0|G^1_0|0\rangle = \epsilon\,(n+1)^{2j-1}\sum_{\gamma K}\begin{bmatrix} J & 0 & J\\ j & j & L\\ K & j & I\end{bmatrix}$$

$$\times\;(j^n aJ;\; j|\}j^{n+1}\gamma K)\;(j^n\beta I;\; j|\}j^{n+1}\gamma K)(-)^{2K}\;\hat{K}. \qquad (7.72)$$

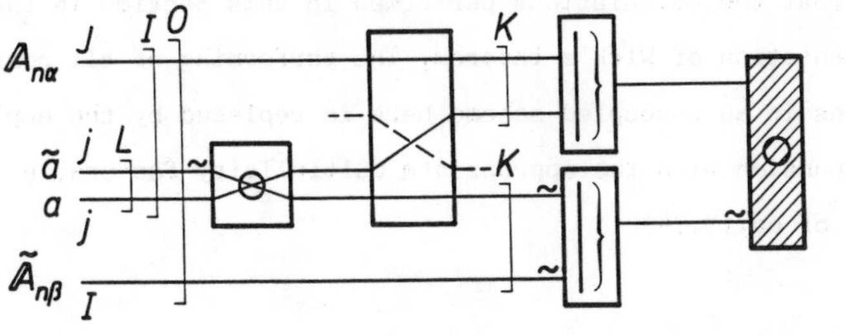

7.14

We also compute the useful case of the mixed 2-body scalar Fock operator

$$\langle 0|G^2_0|0\rangle = \langle 0| \left[\mathbb{A}^{[I]}_{n\alpha} : \left[[\tilde{a}^{[j]}\,\tilde{a}^{[j]}]^{[L]} [a^{[j]}\,a^{[j]}]^{[L]} \right]^{[0]} : \tilde{\mathbb{A}}^{[I]}_{n\beta} \right]^{[0]} |0\rangle . \quad (7.73)$$

The graph 7.15 together with eqs. (7.60) and (7.61) yields after simplification

$$\langle 0|G^2_0|0\rangle = (n+2)(n+1) \sum_{\gamma K} \frac{\hat{K}}{\hat{I}\ \hat{L}} (j^n\alpha I;\ j^2 L||\} j^{n+2}\gamma K)$$

$$\times\ (j^n\beta I;\ j^2 L||\} j^{n+2}\gamma K)(-)^{2K}\ \hat{K}. \quad (7.74)$$

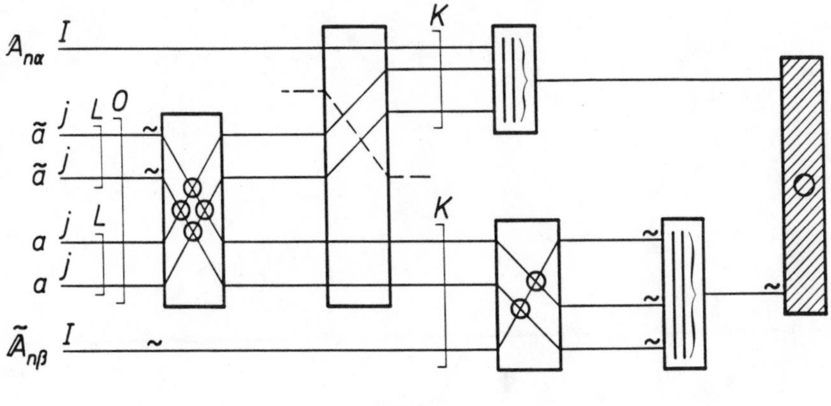

7.15

Note that the calculations performed in this Section in fact amount to the implementation of Wick's theorem. The performing of all possible contractions in an uncoupled scheme here is replaced by the application of the CFP expansion with the appropriate multiplicity factors, e.g. $(n+2)(n+1)$ of eq.(7.74).

The extension to the general case of many-body systems with several shells is straightforward since the many-shell problem can be factorized into the single-shell problems treated above. The needed additional recouplings are readily calculated with the Invariant Graph method. General examples, also including the evaluation of the multiplicity factors, are given in Chapter 7 of the book "Methods in Relativistic Nuclear Physics" (North Holland, 1984, Amsterdam) by Danos, Gillet and Cauvin.

Chapter 8

PARTICLE-HOLE REPRESENTATION

The particle-hole formalism for fermions is greatly simplified by the basic phase convention of this book, which permits to treat on the same footing both particles and holes. We first define the hole representation in tensorial form. Then we show that it is most convenient to re-express the particle-hole operators in terms of particle operators only, however performing expectation values with respect to the particle-hole reference state. This way the elementary invariant matrix elements entering the particle-hole expressions are those already defined in the particle representation, while the particle-hole aspects emerge from the evaluation of the expectation values of the Fock operators. The particle-hole 2-body interaction is calculated as an example.

Reference state and particle-hole operators

In the particle-hole picture for closed shell systems one introduces a reference state, the one in which all fermion states up to the Fermi level are occupied. The reference state is denoted | > and its vector in Fock space is

$$| \rangle = \tilde{a}^{[a]}\, \tilde{a}^{[b]} \ldots \tilde{a}^{[f]} |0\rangle. \tag{8.1}$$

The particle creation operators are written in the lexicographic order.

Single-particle states which belong to the reference state, i.e. up to the Fermi level, are denoted by lower case indices $a,b...,f$, those above the Fermi level by capital indices A, B...

We introduce hole annihilation and creation operators $c^{[a]}_{m_a}$, $\tilde{c}^{[a]}_{m_a}$ acting on the states $a,b,...$ below the Fermi level, defined by their action on the reference state. Thus for the annihilation operators we have

$$c^{[b]}_{m_b}|\,\rangle = 0. \tag{8.2}$$

The creation operators, denoted by the ~ symbol, are then given by

$$\tilde{c}^{[b]}_{m_b} = (-)^{b-m_b}\, c^{[b]+}_{-m_b}, \tag{8.3}$$

in accordance with the basic phase convention as in eq.(7.9). This way both annihilation and creation operators transform as contrastandard tensors.

In terms of the initial particle operators of Chapter 7, these definitions generate the canonical transformation

$$c^{[b]}_{m_b} = (-)^{2b}\, \tilde{a}^{[b]}_{m_b}, \tag{8.4}$$

$$\tilde{c}^{[b]}_{m_b} = a^{[b]}_{m_b}, \tag{8.5}$$

i.e. the commutation relations are preserved, namely

$$\left[c^{[b]}\,\tilde{c}^{[b']}\right]^{[I]}_{M} + (-)^{b+b'-I}\left[\tilde{c}^{[b']}c^{[b]}\right]^{[I]}_{M} = \epsilon\,\hat{b}\,\delta_{bb'}\delta_{I0}. \tag{8.6}$$

In order to complete the representation, the particle annihilation and creation operators for the particle states above the Fermi level are also introduced. They are simply the usual particle operators

$$c^{[B]}_{m_B} = a^{[B]}_{m_B}, \tag{8.7}$$

$$\tilde{c}^{[B]}_{m_B} = \tilde{a}^{[B]}_{m_B}. \tag{8.8}$$

The contraction table with respect to the reference state for the operators c, $\tilde{c}$ yields the two non-zero contractions

$$\langle \mid \left[c^{[b]}\tilde{c}^{[b]}\right]^{[0]} \mid \rangle = \epsilon\ \hat{b}, \tag{8.9}$$

$$\langle \mid \left[c^{[B]}\tilde{c}^{[B]}\right]^{[0]} \mid \rangle = \epsilon\ \hat{B}. \tag{8.10}$$

The key to achieving simple expressions for the particle-hole representation is to express the particle-hole matrix elements in terms of the matrix elements of the usual particle representation. To that end we shall replace the particle-hole operators $\tilde{c}^{[\ ]}, c^{[\ ]}$ by the initial particle operators $\tilde{a}^{[\ ]}, a^{[\ ]}$ in bras, kets and dynamical operators, using the definitions (8.4) and (8.5). The calculation is then carried out introducing for the contractions (8.9) - (8.10) their form written with the initial operators, namely

$$\langle \mid [\tilde{a}^{[b]}a^{[b]}]^{[0]} \mid \rangle = \hat{b}, \tag{8.11}$$

$$\langle \mid \left[a^{[B]}\ \tilde{a}^{[B]}\right]^{[0]} \mid \rangle = (-)^{2B}\ \hat{B}. \tag{8.12}$$

For the graph associated to the hole contraction (8.11), one needs only to introduce the new definition of the end-box of fig.8.1. We note the order of the operators in eqs.(8.11) and (8.12): the tilda line with a hole index enters the end-box at the top in contrast to the case of the particle state contraction end-box, fig.7.3. Also there is no phase ϵ, eq.(7.14). Hence the "hole" contraction box of fig.8.1 is distinguished from a "particle" contraction box, fig.7.3, by the absence of the circle denoting the phase ϵ.

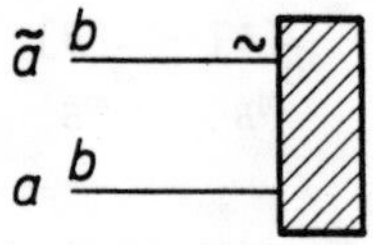

8.1

Particle-hole states

In terms of the particle and hole operators $c^{[\ ]}$, $\tilde{c}^{[\ ]}$ we define the particle-hole ket as

$$|AbI\rangle = (-)^{2I}\ \hat{I}\ \left[W^{[I]}[\tilde{c}^{[A]}\ \tilde{c}^{[b]}]^{[I]}\right]^{[0]}|\ \rangle. \tag{8.13}$$

The corresponding bra is

$$\langle AbI| = \ \hat{I}\ \langle\ |\ \left[\tilde{W}^{[I]}[c^{[A]}c^{[b]}]^{[I]}\right]^{[0]}. \tag{8.14}$$

These two forms ensure normalization to unity. Note the cancellation of the reordering phase ϵ of eqs.(7.29) and (7.33) which is a consequence of the particle-hole representation, in particular eq.(8.4).

Using the canonical transformation of eqs.(8.4)-(8.5) and (8.7)-(8.8) all particle-hole expressions from now on will be written in terms of the original Fock operators $a^{[\ ]}$, $\tilde{a}^{[\ ]}$, as explained above. Thus the particle-hole ket, eq.(8.13), becomes

$$|AbI\rangle = (-)^{2I}\ \hat{I}\ \left[W^{[I]}[\tilde{a}^{[A]}a^{[b]}]^{[I]}\right]^{[0]}|\ \rangle, \tag{8.15}$$

and the bra, eq.(8.14)

$$\langle AbI| = (-)^{2b}\ \ \hat{I}\ \langle\ |\ \left[\tilde{W}^{[I]}[a^{[A]}\tilde{a}^{[b]}]^{[I]}\right]^{[0]}. \tag{8.16}$$

These forms permit to use directly the expressions for any operator Ω_λ written in terms of the usual particle representation as given in Chapter

7, as we now show.

Particle-hole one-body matrix elements

The one-body multipole operator Ω_λ, eq(7.41), has four distinct terms since the two particle indices i,k each have two ranges of values, viz, above and below the Fermi level. Thus eq.(7.41) is directly developed as

$$\begin{aligned}\Omega_\lambda = &\sum_{AB} \left[\varphi^{[A]}\middle|\Omega^{[\lambda]}\middle|\varphi^{[B]}\right] :[\tilde{a}^{[A]}\omega^{[\lambda]}a^{[B]}]^{[0]}: \\ &+ \sum_{ab} \left[\varphi^{[a]}\middle|\Omega^{[\lambda]}\middle|\varphi^{[b]}\right] :[\tilde{a}^{[a]}\omega^{[\lambda]}a^{[b]}]^{[0]}: \\ &+ \sum_{Ab} \left[\varphi^{[A]}\middle|\Omega^{[\lambda]}\middle|\varphi^{[b]}\right] :[\tilde{a}^{[A]}\omega^{[\lambda]}a^{[b]}]^{[0]}: \\ &+ \sum_{aB} \left[\varphi^{[a]}\middle|\Omega^{[\lambda]}\middle|\varphi^{[B]}\right] :[\tilde{a}^{[a]}\omega^{[\lambda]}a^{[B]}]^{[0]}: .\end{aligned} \tag{8.17}$$

These four terms correspond respectively to the four processes : particle scattering, hole scattering, particle-hole pair creation and particle-hole pair annihilation. Note that the order of the particle and hole indices in the matrix elements and in the normal products of the Fock operators is identical.

In performing a calculation, the matrix of this operator will be written with the bras and kets of the particle-hole representation expressed in terms of the initial operators $a^{[\ ]}$, $\tilde{a}^{[\ ]}$, as in eqs. (8.15) and (8.16). The contractions are to be performed with respect to the reference state. Their values are given in eqs.(8.11) and (8.12).

For example the matrix element of the pair creation process is given by :

$$
\begin{aligned}
M &= \langle AbI |\, \Omega_{\lambda} \,| \rangle \\
&= [\varphi^{[A]} | \Omega^{[\lambda]} | \varphi^{[b]}] (-)^{2b} \, \hat{I} \\
&\quad \times \langle \,|\, \left[\tilde{W}^{[I]} [a^{[A]} \, \tilde{a}^{[b]}]^{[I]} \right]^{[0]} : \left[\tilde{a}^{[A]} \omega^{[\lambda]} a^{[b]} \right]^{[0]} : \,| \rangle \\
&= [\varphi^{[A]} | \Omega^{[\lambda]} | \varphi^{[b]}] \langle \,| G | \rangle .
\end{aligned}
\tag{8.18}
$$

As always the result appears as the product of an invariant matrix element of $\Omega^{[\lambda]}$ and of an expectation value $\langle \,|G|\, \rangle$.

The factor $\langle \,|G|\, \rangle$ is evaluated in fig.8.2, which yields with the selection rule $\lambda = I$

$$
\begin{aligned}
\langle \,|G|\, \rangle &= (-)^{2b} \, \hat{I} \, (-)^{A+I-b} \begin{bmatrix} I & I & 0 \\ I & I & 0 \\ 0 & 0 & 0 \end{bmatrix} \begin{bmatrix} A & b & I \\ A & b & I \\ 0 & 0 & 0 \end{bmatrix} \epsilon^2 \, \hat{A} \, \hat{b} \, [W^{[I]} \omega^{[I]}]^{[0]} \\
&= (-)^{b+A+I} [W^{[I]} \omega^{[I]}]^{[0]} .
\end{aligned}
\tag{8.19}
$$

Note the difference between the particle contraction and the hole contraction end-boxes. The particle-hole aspect is here contained in the phase.

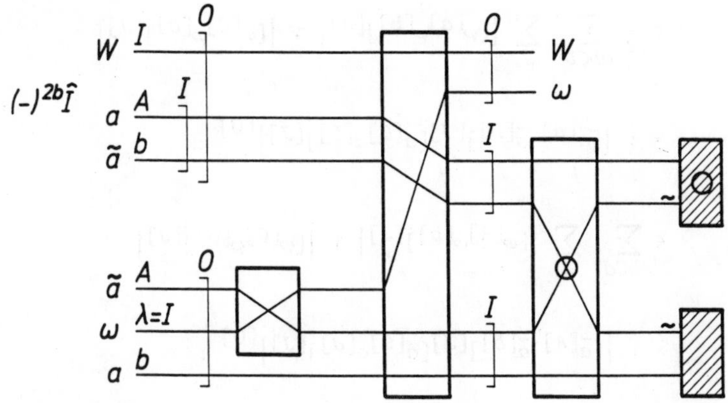

8.2

Particle-hole two-body matrix elements

We consider as the next example the two-body scalar operator V, eq.(7.38). There are now sixteen different terms which arise from giving each particle index its range of variation either as a particle, i=A, or as a hole, i=b. The sixteen terms contribute each to different processes, as defined by the bras and kets. Of these sixteen terms we write out only four

$$V = \epsilon \, \frac{1}{2} \sum_{ABCD} \sum_{J} \left[[\varphi^{[A]} \varphi^{[B]}]^{[J]} \right| \mathcal{V} \left| [\varphi^{[C]} \varphi^{[D]}]^{[J]} \right]$$

$$\times : \left[[\tilde{a}^{[A]} \, \tilde{a}^{[B]}]^{[J]} [a^{[C]} a^{[D]}]^{[J]} \right]^{[0]} :$$

$$+ \epsilon \sum_{aBCD} \sum_{J} \left[[\varphi^{[a]} \varphi^{[B]}]^{[J]} \right| \mathcal{V} \left| [\varphi^{[C]} \varphi^{[D]}]^{[J]} \right]$$

$$\times : \left[[\tilde{a}^{[a]} \, \tilde{a}^{[B]}]^{[J]} [a^{[C]} a^{[D]}]^{[J]} \right]^{[0]} :$$

$$+ \epsilon \frac{1}{2} \sum_{abCD} \sum_{J} [[\varphi^{[a]}\varphi^{[b]}]^{[J]}| \mathcal{V} |[\varphi^{[C]}\varphi^{[D]}]^{[J]}]$$

$$\times : [[\tilde{a}^{[a]} \tilde{a}^{[b]}]^{[J]}[a^{[C]}a^{[D]}]^{[J]}]^{[0]} :$$

$$+ \epsilon \sum_{AbCd} \sum_{J} [[\varphi^{[A]}\varphi^{[b]}]^{[J]}| \mathcal{V} |[\varphi^{[C]}\varphi^{[d]}]^{[J]}]$$

$$\times : [[\tilde{a}^{[A]} \tilde{a}^{[b]}]^{[J]}[a^{[C]}a^{[d]}]^{[J]}]^{[0]} :$$

$$+ 12 \ \textit{terms}\ldots .$$

(8.20)

The phase $\epsilon = (-)$ results from our choice for ordering the annihilation operators, as explained with reference to eq.(7.36). The sums over the state indices are independent. Hence the factors 1/2 for the terms where there is double counting. The four terms given explicitly in eq.(8.20) contribute respectively to 2-particle scattering, particle scattering with pair annihilation, 2-particle 2-hole annihilation (vacuum fluctuation), particle-hole scattering.

We give now as an example the calculation of the matrix element for particle-hole scattering. The initial and final states of the process are represented by the particle-hole ket and bra respectively

$$|Ab;I\rangle = (-)^{2I} \hat{I} \left[W^{[I]}[\tilde{a}^{[A]}a^{[b]}]^{[I]}\right]^{[0]}| \ \rangle,$$

$$\langle Cd;I| = (-)^{2d} \ \hat{I} \langle \ | \left[\tilde{W}^{[I]}[a^{[C]}\tilde{a}^{[d]}]^{[I]}\right]^{[0]},$$

(8.21)

We select among the 16 terms of the operator (8.20) those which yield non-zero contributions, of which there are two. The first one, conventionally called the direct term, is

$$M = \epsilon(-)^{2I+2d}\, \hat{I}^2 \sum_J \; \langle \,|\, \left[\tilde{W}^{[I]}[a^{[C]}\tilde{a}^{[d]}]^{[I]}\right]^{[0]}$$

$$\times \; : \left[\tilde{a}^{[b]}\, \tilde{a}^{[C]}]^{[J]}[a^{[A]}a^{[d]}]^{[J]}\right]^{[0]} \; : \left[W^{[I]}[\tilde{a}^{[A]}a^{[b]}]^{[I]}\right]^{[0]} \;|\, \rangle$$

$$\times \left[[\varphi^{[b]}(1)\varphi^{[C]}(2)]^{[J]}\right| \mathcal{V}(1,2) \left|[\varphi^{[A]}(1)\varphi^{[d]}(2)]^{[J]}\right]. \tag{8.22}$$

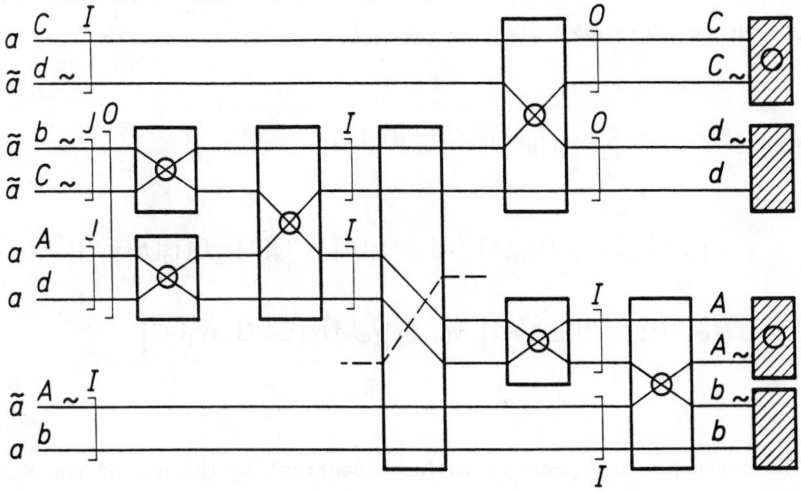

8.3

The expectation value is computed in the graph 8.3. For simplicity we have omitted the first part of the graph which as usual represents the separation of the amplitudes W and the performing of their overlaps, it contributes the usual factor $\frac{1}{\hat{I}}$. We note the distribution of the particle indices in the invariant matrix element, characteristic of the particle-hole scheme. The reader will also note the positions of the tilda lines entering the contraction boxes : at the top for particles, at the bottom for holes, according to eqs.(8.11), (8.12) respectively. From the graph we read off the result which, after simplification, is

$$M = \frac{1}{\hat{I}} (-)^{C+d-I} \sum_J \begin{bmatrix} C & b & J \\ d & A & J \\ I & I & 0 \end{bmatrix}$$

$$\times \left[[\varphi^{[b]}(1)\varphi^{[C]}(2)]^{[J]} |\ \mathcal{V}(1,2)\ | [\varphi^{[A]}(1)\varphi^{[d]}(2)]^{[J]} \right]. \tag{8.23}$$

The other particle-hole scattering matrix element, conventionally called the exchange term, writes, again paying atention to the distribution of the indices associated with the process,

$$M = \epsilon(-)^{2I+2d}\ \hat{I}^2 \sum_J \langle\ |\left[W^{[I]}[a^{[C]}\tilde{a}^{[d]}]^{[I]}\right]$$

$$\times :\ [\tilde{a}^{[C]}\ \tilde{a}^{[b]}]^{[J]}[a^{[A]}a^{[d]}]^{[J]}]^{[0]}\ :\ \left[W^{[I]}[\tilde{a}^{[A]}a^{[b]}]^{[I]}\right]^{[0]} |\ \rangle$$

$$\times \left[[\varphi^{[C]}(1)\varphi^{[b]}(2)]^{[J]} \right| \mathcal{V}(1,2) \left| [\varphi^{[A]}(1)\varphi^{[d]}(2)]^{[J]} \right]. \tag{8.24}$$

The corresponding graph is in fact identical to the one of the first process, fig.8.3, after performing in the normal product of the operator the permutation of fig.8.4. Thus the result

$$M = \frac{1}{\hat{I}} (-)^{b+d-I} \sum_J (-)^J \begin{bmatrix} C & b & J \\ d & A & J \\ I & I & 0 \end{bmatrix}$$

$$\times\ [[\varphi^{[C]}(1)\varphi^{[b]}(2)]^{[J]} |\ \mathcal{V}(1,2)\ | [\varphi^{[A]}(1)\varphi^{[d]}(2)]^{[J]}]. \tag{8.25}$$

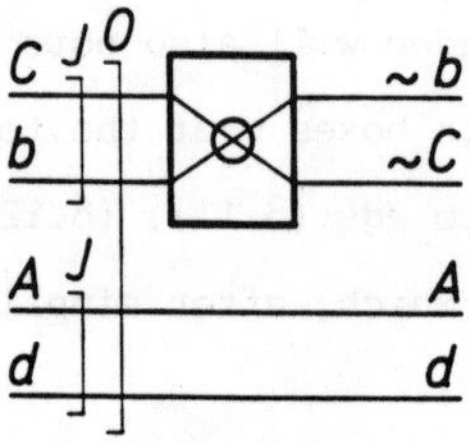

8.4

Chapter 9

TRANSITION PROBABILITIES AND ANGULAR DISTRIBUTIONS

The description of transitions and reactions, including angular distributions and polarizations, constitutes the most complicated application of angular momentum calculus. The Invariant Graph method is particularly useful in such involved situations because it gives a complete overview over the full process while at the same time leading directly to the final algebraic expression. In these applications the use of the density matrix formalism, which is requisite for impure states, allows even for pure states a compact and efficient description of the physics. This formalism is well adapted to the invariant graph method and we use it throughout.

The examples we treat are very general. Owing to the large number of physical observables, they of necessity lead to lengthy expressions, but here without extraneous summations. Considerable simplifications arise when some observables are not measured, and we give the expressions for some such cases.

Transitions

The probability for a transition from state Ψ_I to state Φ_J induced by an operator T_λ is proportional to

$$P = \left|\langle\Phi_J|T_\lambda|\Psi_I\rangle\right|^2, \tag{9.1}$$

which can be re-written as

$$P = \langle\Psi_I|T_\lambda^\dagger|\Phi_J\rangle\ \langle\Phi_J|T_\lambda|\Psi_I\rangle. \tag{9.2}$$

We introduce the notation

$$\Omega\ = T_\lambda^\dagger|\Phi_J\rangle\ \langle\Phi_J|T_\lambda. \tag{9.3}$$

Then P can be written as an expectation value for the state Ψ_I

$$P = \langle\Psi_I|\Omega\ |\Psi_I\rangle. \tag{9.4}$$

Expanding Ω into multipoles, which is always possible,

$$\Omega = \sum_K\ \hat{K}\ [\omega^{[K]}\Omega^{[K]}]^{[0]}, \tag{9.5}$$

we obtain, from fig.4.5 and eq.(4.11),

$$P = \sum_K\ \left[\Psi^{[I]}\middle|\Omega^{[K]}\middle|\Psi^{[I]}\right]\ [\tilde{W}^{[I]}\omega^{[K]}W^{[I]}]^{[0]}. \tag{9.6}$$

The transition probability thus separates into an invariant matrix element describing the dynamics and and an invariant product of the amplitudes associated with the experimental set-up.

We may re-write the expression (9.6) in terms of the density matrix $\rho^{[K]}$

$$P = \sum_K\ (-)^{2I}\ \left[\Psi^{[I]}\middle|\Omega^{[K]}\middle|\Psi^{[I]}\right]\ [\rho_I^{[K]}\omega^{[K]}]^{[0]}, \tag{9.7}$$

where the definition of $\rho^{[K]}$ arises by cyclic permutation in the invariant triple product

$$[\tilde{W}^{[I]}\omega^{[K]}W^{[I]}]^{[0]} = (-)^{2I}\ [W^{[I]}\tilde{W}^{[I]}\omega^{[K]}]^{[0]}, \tag{9.8}$$

hence

$$\rho_I^{[K]} = [W^{[I]}\tilde{W}^{[I]}]^{[K]}. \tag{9.9}$$

Even though calculations for pure states can be performed using the amplitudes $W_M^{[I]}$ themselves, the density matrices permit a very convenient and compact notation for transition probabilities.

The density matrices as defined here, eq.(9.9), transform under rotation as contrastandard tensors,

$$\rho'^{[K]}_{IM} = \sum_{M'} \rho^{[K]}_{IM'}\ \mathscr{D}^K_{M'M}(\alpha\beta\gamma). \tag{9.10}$$

For pure states all multipolarities between $K = 0$ and $K = 2I$ exist. In contrast, for a fully unpolarized state only the component with $K = 0$ of the density matrix exists. In general, for impure states which must be described directly by a density matrix, i.e. where the factorization according to eq.(9.9) is not valid, the expression of eq.(9.7) for P provides the starting point for all further calculations.

For pure and impure states we have

$$\rho_I^{[0]} = (-)^{2I}\ \frac{1}{\hat{I}}, \tag{9.11}$$

which reflects the normalization of the state.

The expression (9.7) is deceptively simple since all complications are hidden in the matrix element of the operator Ω, eq.(9.3). We now compute in detail the most general form of P. Generalizing T to a mixture of multipoles

$$T = \sum_{\lambda} \hat{\lambda}\ [t^{[\lambda]}T^{[\lambda]}]^{[0]}, \tag{9.12}$$

we define the operator density matrix

$$\rho^{[\Lambda]}_{\lambda\lambda'} = [t^{[\lambda]}\tilde{t}^{[\lambda']}]^{[\Lambda]}, \tag{9.13}$$

and the initial and final state density matrices $\rho^{[K]}_{i}$ and $\rho^{[L]}_{f}$ which in the case of pure states write respectively

$$\rho^{[K]}_{i} = [W^{[I]}_{i}\tilde{W}^{[I]}_{i}]^{[K]}, \tag{9.14}$$

and

$$\rho^{[L]}_{f} = [W^{[J]}_{f}\tilde{W}^{[J]}_{f}]^{[L]}. \tag{9.15}$$

We thus have to evaluate the expression

$$P_{fi} = \left|\langle\Phi_{fJ}|T|\Psi_{iI}\rangle\right|^2$$

$$= \sum_{\lambda\lambda'} \hat{J}^2\hat{I}^2\hat{\lambda}\ \hat{\lambda}'\left[[W^{[I]}_{i}\Psi^{[I]}_{i}]^{[0]}\Big|[\tilde{t}^{[\lambda']}\ \tilde{T}^{[\lambda']}]^{[0]}\ \Big|[W^{[J]}_{f}\Phi^{[J]}_{f}]^{[0]}\right]$$

$$\times\left[[W^{[J]}_{f}\Phi^{[J]}_{f}]^{[0]}\Big|\ [t^{[\lambda]}\ T^{[\lambda]}]^{[0]}\ \Big|[W^{[I]}_{i}\Psi^{[I]}_{i}]^{[0]}\right]. \tag{9.16}$$

In drawing the recoupling graph, fig.9.1, we use again the possibility to freely place commuting angular momentum zero quantities as dictated by convenience. Hence we adopt the order of coupled tensors shown on the left hand side of fig.9.1 to obtain in the simplest manner the definitions of eq.(9.9) for the density matrices.

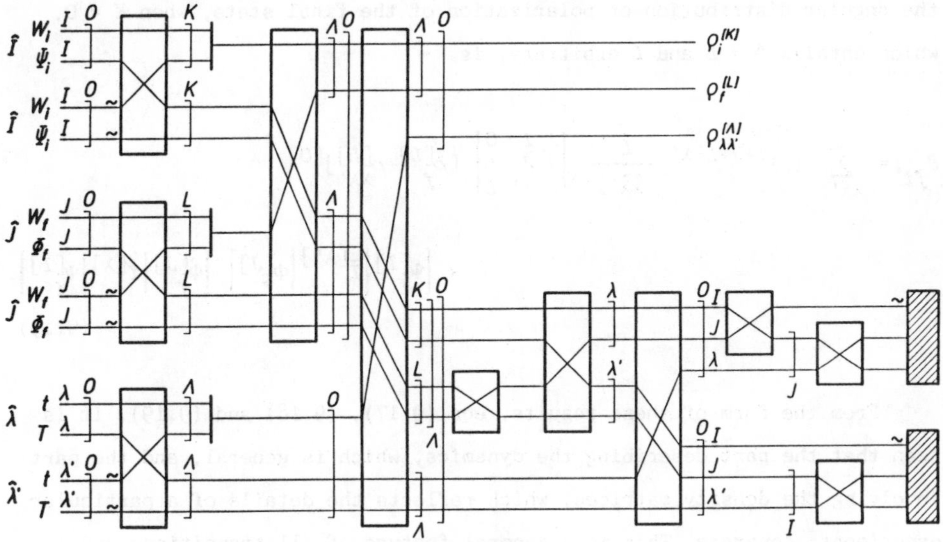

9.1

The algebraic expression read off the graph becomes after carrying out the simplifications

$$P_{fi} = \sum_{\lambda\lambda' KL\Lambda} (-)^{I+\lambda'-J-L} \frac{\hat{\Lambda}}{\hat{\lambda}\hat{\lambda}'} \begin{bmatrix} I & I & K \\ J & J & L \\ \lambda & \lambda' & \Lambda \end{bmatrix} [\rho_i^{[K]}\ \rho_f^{[L]}\ \rho_{\lambda\lambda'}^{[\Lambda]}]^{[0]} \left[\Psi_i^{[I]} \middle| \tilde{T}^{[\lambda']} \middle| \Phi_f^{[J]}\right]$$

$$\times \left[\Phi_f^{[J]} \middle| T^{[\lambda]} \middle| \Psi_i^{[I]}\right]. \tag{9.17}$$

The transition probability takes on a very simple form for an unpolarized initial state, for which $K = 0$ and $\rho_i^{[0]} = (-)^{2I}/\hat{I}$, and when one does not measure the angular distribution and polarization of the final state, in which case $L = 0$ and $\rho_f^{[0]} = (-)^{2J}/\hat{J}$. The result then is

$$P_{fi} = \sum_{\lambda} (-)^{I+\lambda-J} \frac{1}{\hat{I}^2\ \hat{J}^2\ \hat{\lambda}^2} \left[\Psi_i^{[I]} \middle| \tilde{T}^{[\lambda]} \middle| \Phi_f^{[J]}\right] \left[\Phi_f^{[J]} \middle| T^{[\lambda]} \middle| \Psi_i^{[I]}\right]. \tag{9.18}$$

This expression of course is the total transition probability to the final state $\Phi_f^{[J]}$.

The case for an unpolarized initial state, but with the measurement of the angular distribution or polarization of the final state, when $K = 0$, which entails $\Lambda = L$ and L arbitrary, is

$$P_{fi} = \sum_{\lambda\lambda' L} (-)^{I+J+L-\lambda'} \frac{\hat{L}}{\hat{\lambda}\hat{\lambda}'\hat{I}} \begin{bmatrix} I & I & 0 \\ J & J & L \\ \lambda & \lambda' & L \end{bmatrix} [\rho_f^{[L]}\ \rho_{\lambda\lambda'}^{[L]}]^{[0]}$$

$$\times \left[\Psi_i^{[I]}\middle|\tilde{T}^{[\lambda']}\middle|\Phi_f^{[J]}\right] \left[\Phi_f^{[J]}\middle|T^{[\lambda]}\middle|\Psi_i^{[I]}\right]. \tag{9.19}$$

From the form of these results, eqs.(9.17), (9.18) and (9.19), it is seen that the part describing the dynamics, which is general, and the part involving the density matrices, which reflects the details of a particular experiment, separate. This is a general feature of all transitions and reactions.

Density matrix of a beam

Before discussing the treatment of reactions, we devote this section to the construction of different density matrices of a beam. The examples we give should cover most common situations.

At infinity a beam is described by a plane wave. It has the multipole expansion

$$e^{i\vec{k}\cdot\vec{r}} = 4\pi \sum_{\ell} i^{\ell}\, \hat{\ell}\, j_{\ell}(kr)\ [Y^{[\ell]}(\hat{k})Y^{[\ell]}(\hat{r})]^{[0]}. \tag{9.20}$$

To make contact with our usual notation we note that the asignment for the normalized amplitudes

$$W_m^{[\ell]}(\hat{k}) = \frac{1}{\hat{\ell}} Y_m^{[\ell]}(\hat{k}), \tag{9.21}$$

is appropriate as our spherical harmonics fulfill the required phase convention and the normalisation is indeed

$$\sum_m \int d^2\hat{k} \left|W_m^{[\ell]}(\hat{k})\right|^2 = 1. \tag{9.22}$$

Introducing the notation

$$\varphi_{km}^{[\ell]}(\hat{r}) = 4\pi \ \hat{\ell}\ j_\ell(kr) Y_m^{[\ell]}(\hat{r}), \tag{9.23}$$

we can write

$$e^{i\vec{k}\cdot\vec{r}} = \sum_\ell i^\ell \ \hat{\ell}\ [W^{[\ell]}(\hat{k})\ \varphi_k^{[\ell]}(\hat{r})]^{[0]}. \tag{9.24}$$

Note that we have kept the plane wave phase i^ℓ as an explicit factor since it plays an important role in the interference between different multipolarities. The beam density matrix thus is

$$\rho_{\ell\ell' M}^{[L]}(\hat{k}) = [W^{[\ell]}(\hat{k})\ \tilde{W}^{[\ell']}(\hat{k})]_M^{[L]} = \frac{1}{\hat{\ell}\ \hat{\ell}'}\ [Y^{[\ell]}(\hat{k}) Y^{[\ell']}(\hat{k})]_M^{[L]}. \tag{9.25}$$

This expression can be evaluated further ; from fig.9.2, we obtain

$$\rho_{\ell\ell' M}^{[L]}(\hat{k}) = \frac{1}{\hat{\ell}\ \hat{\ell}'\hat{L}}\ [\ell|\ell'|L]\ Y_M^{[L]}(\hat{k}), \tag{9.26}$$

introducing the notation of eq.(4.20) for the invariant multipole matrix element.

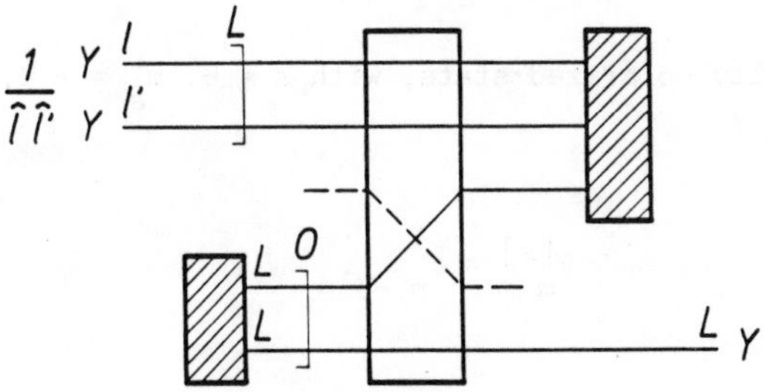

9.2

The case $M = 0$ reduces to

$$\rho^{[L]}_{\ell\ell'0}(\hat{k}) = (-i)^L \frac{1}{\hat{\ell}\,\hat{\ell}'} \frac{1}{\sqrt{4\pi}} [\ell|\ell'|L]\, P_L(\cos\theta). \tag{9.27}$$

It further simplifies for the case of the beam being parallel to the Oz axis. Then $\hat{k} = k_z$ and $\theta = \varphi = 0$, $P_L = 1$,

$$\rho^{[L]}_{\ell\ell'0}(\hat{k}_z) = (-i)^L \frac{1}{\sqrt{4\pi}} [\ell|\ell'|L], \tag{9.28}$$

which for $L = 0$ is

$$\rho^{[0]}_{\ell\ell'}(\hat{k}_z) = \frac{1}{\sqrt{4\pi}}\ \frac{1}{\hat{\ell}}\,\delta_{\ell\ell'}\ . \tag{9.29}$$

We now consider the case of a plane wave beam of polarized particles. In the notation of Chapter 3 for the polarized state and of eq.(9.24) for the plane wave, we have

$$\hat{s}\,[W^{[s]}\chi^{[s]}]^{[0]} e^{i\vec{k}\cdot\vec{r}} = \sum_{\ell} i^{\ell}\,\hat{s}\,\hat{\ell}\,[W^{[s]}\chi^{[s]}]^{[0]}[W^{[\ell]}(\hat{k})\varphi^{[\ell]}(\vec{r})]^{[0]}. \tag{9.30}$$

For example for a fully polarized state, with $s = ½$, $m_s = +½$, the amplitudes are

$$-W^{[s]}_m = \delta_{m,-½}. \tag{9.31}$$

The spin density matrix

$$\rho_M^{[S]} = [W^{[s]}\tilde{W}^{[s]}]_M^{[S]} = \sum_{mm'} (ssmm'|SM)\ W_m^{[s]}\tilde{W}_{m'}^{[s]}, \tag{9.32}$$

has in this case only the two components

$$\rho_0^{[0]} = (ss\text{-}\tfrac{1}{2}\ \tfrac{1}{2}|00) = -\frac{1}{\hat{s}},$$

$$\rho_0^{[1]} = (ss\text{-}\tfrac{1}{2}\ \tfrac{1}{2}|10) = \frac{1}{\hat{s}}\ . \tag{9.33}$$

Next we consider the case of a beam of spin 1 particles with helicity + 1, i.e. with polarization $m = +1$. Then

$$W_m^{[1]} = -i\delta_{m,-1},$$

$$\tilde{W}_m^{[1]} = i\delta_{m,1}. \tag{9.34}$$

Herewith we have from

$$\rho_0^{[S]} = \sum_{mm'} (11mm'|S0)\ W_m^{[1]}\ \tilde{W}_{m'}^{[1]}, \tag{9.35}$$

the values

$$\rho^{[0]} = \frac{1}{\sqrt{3}}\ ,$$

$$\rho_0^{[1]} = -\frac{1}{\sqrt{2}}\ ,$$

$$\rho_0^{[2]} = \frac{1}{\sqrt{6}}\ . \tag{9.36}$$

Again all other components of the density matrices vanish. For an unpolarized beam only $\rho^{[0]}$ survives.

As a somewhat more involved example we consider the cases of linear polarization. For polarization along the Oy-axis we have

$$\vec{W}\cdot\vec{e} = W_y e_y = \hat{1}\ [W^{[1]} e^{[1]}]^{[0]}, \tag{9.37}$$

with, eq. (3.31) and with $W_y = 1,\ W_x = W_z = 0,$

$$W^{[1]}_m = \tilde{W}^{[1]}_m = \frac{1}{\sqrt{2}}\,\delta_{|m|,1}. \tag{9.38}$$

Similarly for polarization along the Ox-axis we have

$$\vec{W}\cdot\vec{e} = W_x e_x = \hat{1}\ [W^{[1]} e^{[1]}]^{[0]}, \tag{9.39}$$

with

$$W^{[1]}_m = \tilde{W}^{[1]}_m = i\,\frac{1}{\sqrt{2}}\,\delta_{|m|,1}\,m. \tag{9.40}$$

The corresponding density matrices are almost the same for both these cases. For the polarization along the Oy-axis we find

$$\rho^{[0]} = \frac{1}{\sqrt{3}},$$

$$\rho^{[1]}_M = 0,$$

$$\rho^{[2]}_{\pm 2} = \frac{1}{2}, \qquad \rho^{[2]}_{\pm 1} = 0, \qquad \rho^{[2]}_0 = \frac{1}{\sqrt{6}}. \tag{9.41}$$

The only change for the polarization along the Ox-axis is that the elements

$\rho^{[2]}_{\pm 2}$ are negative.

Reactions

We consider the case of a reaction between fully specified channels, the initial channel α and the final channel β, going through all the non-observed intermediate states γ permitted by the conservation and selection rules. This is the most general situation which contains all the aspects of angular distribution and polarization calculations. When some observables are not measured the simplified expressions will also be given.

The channel α is defined by the target state *A* and the projectile *a*. For the final state we consider a residual system *B* and an emitted reaction product *b*. The intermediate states — which here are considered as not observed — constitute the system *C*, with quantum numbers γ. Thus we have

$$(A+a)_{\alpha} \longrightarrow (C)_{\gamma} \longrightarrow (B+b)_{\beta}. \tag{9.42}$$

The reaction cross section then is proportional to

$$P_{\alpha\beta} = \left| \sum_{\gamma} \langle\beta|\, \mathcal{V}\, |\gamma\rangle \langle\gamma|\, \mathcal{U}\, |\alpha\rangle \right|^2, \tag{9.43}$$

or, in detail

$$P_{\alpha\beta} = \sum_{\gamma_1\gamma_2} \langle\alpha|\, \mathcal{U}^{\dagger}\, |\gamma_2\rangle \langle\gamma_2|\, \mathcal{V}^{\dagger}\, |\beta\rangle \langle\beta|\, \mathcal{V}\, |\gamma_1\rangle \langle\gamma_1|\, \mathcal{U}\, |\alpha\rangle. \tag{9.44}$$

Here $\langle\gamma|\mathcal{U}|\alpha\rangle$ etc., are the reaction matrix elements between the different states and are given by the reaction theory one choses to employ. They contain both the dynamics of the reaction and the description of the particular experimental set-up : preparation of the initial state (for example, polarized beams and targets), prescription (measurement) for the final state leading to angular distributions, polarizations etc. As always, we first separate these two aspects, i.e. the dynamics and the experimental set up. This separation is shown in the recoupling graph fig.9.3a.

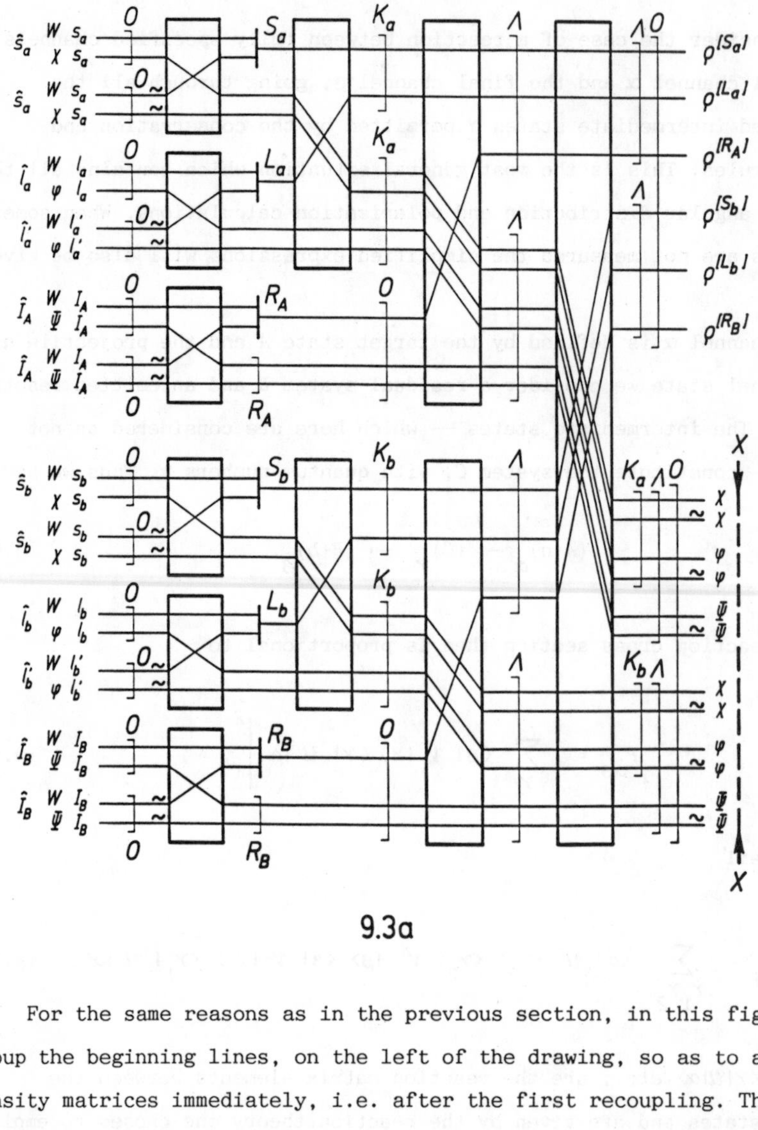

9.3a

For the same reasons as in the previous section, in this figure we group the beginning lines, on the left of the drawing, so as to achieve the density matrices immediately, i.e. after the first recoupling. Then the purpose of the remaining part of the graph is to separate the density matrices from the tensors associated with the dynamics of the reaction. This is reached at the right hand side of the figure. The heavy vertical dashed line X-X indicates the dynamical part, which will be transferred to fig.9.3b for the final completion of the calculation. The reader may verify that the resulting algebraic expression of fig.9.3a is given by

$$P_{\alpha\beta}(\vec{k}_a,\vec{k}_b,\rho's) = (F)(G), \tag{9.45}$$

where (F) is the algebraic expression for the graph 9.3a which after simplification yields

$$(F) = 1, \tag{9.46}$$

and where the function (G) is given by

$$(G) = \Big[\big[[\rho^{[S_a]}\rho^{[L_a]}]^{[K_a]}\ \rho^{[R_A]}\big]^{[\Lambda]}\big[[\rho^{[S_b]}\rho^{[L_b]}]^{[K_b]}\rho^{[R_B]}\big]^{[\Lambda]}\Big]^{[0]}$$

$$\times\Bigg[\Big[\big[[\chi^{[s_a]}\tilde{\chi}^{[s_a]}]^{[S_a]}[\varphi^{[\ell_a]}\tilde{\varphi}^{[\ell_a']}]^{[L_a]}\big]^{[K_a]}\ [\Psi^{[I_A]}\tilde{\Psi}^{[I_A]}]^{[R_A]}\Big]^{[\Lambda]}$$

$$\times\Big[\big[[\chi^{[s_b]}\tilde{\chi}^{[s_b]}]^{[S_b]}[\varphi^{[\ell_b]}\tilde{\varphi}^{[\ell_b']}]^{[L_b]}\big]^{[K_b]}\ [\Psi^{[I_B]}\tilde{\Psi}^{[I_B]}]^{[R_B]}\Big]^{[\Lambda]}\Bigg]^{[0]} . \tag{9.47}$$

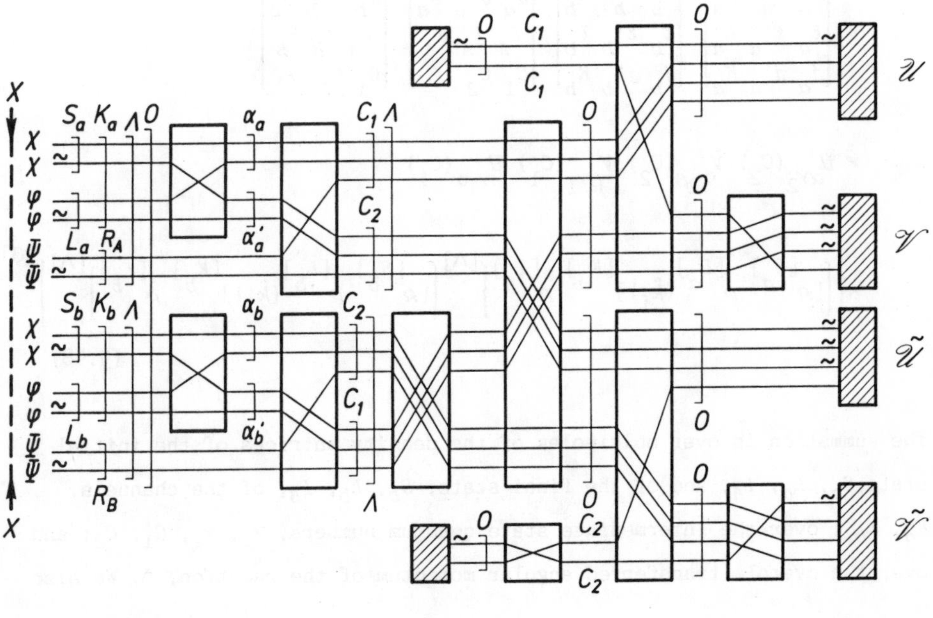

9.3b

The calculation is completed in fig.9.3b. The lines which make up a channel are coupled to the total angular momentum C_1, C_2 of the intermediate states defined by the quantum numbers γ_1 and γ_2. Finally each of these coupled channel lines, together with the appropriate lines representing the inserted intermediate states γ_1, γ_2 are terminated in the respective end-boxes representing the invariant reaction amplitudes denoted $\mathcal{U}_{ij}(C)$ and $\mathcal{V}_{ij}(C)$.The values of these $\mathcal{U}$ and $\mathcal{V}$ boxes must have the normalization of an invariant matrix element, i.e. $\mathcal{U}_{ij}(C) = [i|\mathcal{U}^{[C]}|j]$. They are obtained from reaction theory. One must be careful to always have the tilda channel entering at the top of the end-boxes, and to combine the proper initial and final channels, as given in eq.(9.44).

Thus the expression for the complete reaction probability read off from fig. 9.3b is after simplification

$$P_{\alpha\beta}(\hat{k}_a,\hat{k}_b,\rho's) = \sum (-)^{C_1+C_2+\Lambda} \frac{\hat{\Lambda}}{\hat{C}_1^2\hat{C}_2^2}$$

$$\times \begin{bmatrix} s_a & s_a & S_a \\ \ell_a & \ell'_a & L_a \\ j_a & j'_a & K_a \end{bmatrix} \begin{bmatrix} s_b & s_b & S_b \\ \ell_b & \ell'_b & L_b \\ j_b & j'_b & K_b \end{bmatrix} \begin{bmatrix} j_a & j'_a & K_a \\ I_A & I_A & R_A \\ C_1 & C_2 & \Lambda \end{bmatrix} \begin{bmatrix} j_b & j'_b & K_b \\ I_B & I_B & R_B \\ C_1 & C_2 & \Lambda \end{bmatrix}$$

$$\times \tilde{\mathcal{U}}_{\alpha\gamma_2}(C_2)\, \tilde{\mathcal{V}}_{\gamma_2\beta}(C_2)\, \mathcal{V}_{\beta\gamma_1}(C_1)\, \mathcal{U}_{\gamma_1\alpha}(C_1)$$

$$\times \left[\left[[\rho^{[S_a]}\, \rho^{[L_a]}(\hat{k}_1)]^{[K_a]}\, \rho^{[I_A]} \right]^{[\Lambda]} \left[[\rho^{[S_b]}\, \rho^{[L_b]}(\hat{k}_2)]^{[K_b]}\, \rho^{[I_B]} \right]^{[\Lambda]} \right]^{[0]} . \tag{9.48}$$

The summation is over multipoles of the density matrices of the initial state, S_a, L_a, I_A, and of the final state, S_b, L_b, I_B, of the channels, K_a, K_b ; over the intermediate state quantum numbers, γ_1, γ_2, C_1, C_2; and over the overall transferred angular momentum of the reaction, Λ. We also have summations over the angular momenta of the multipole decompositions of the initial and final beams, ℓ_a, ℓ_a', ℓ_b, ℓ_b' and the spin-orbit total angular momenta j_a, j_a', j_b, j_b'.

Note that all these summations involve quantum numbers which are attached to physically significant quantities, i.e. reaction amplitudes and density matrices. In this case by judicious choice in drawing the recoupling graph all dummy summations could be avoided.

In this formulation a measurement on the final channel, for example of the polarization of the outgoing particles, is in fact specified by the density matrix for that quantity. This specification of the measured final channels by their density matrices is equivalent to the introduction of projection operators specifying the measuring apparatus. The use of density matrices throughout allows a more compact form for the final expression,

We now list standard cases in which some of the observables are not measured, which lead to simplification of the general expression (9.48).

As a first exemple we compute the angular distribution for the reaction (9.42) without polarizations and for a unique intermediate state Υ. In that case $S_a = 0$, $S_b = 0$, $R_A = 0$, $R_B = 0$, which then enforces $\Lambda = L_a = L_b = K_a = K_b$ and $C_1 = C_2 = C$. We also assume that the incoming beam is along the 0z direction. Substituting the values

$$\rho^{[0]}_{ss} = \frac{(-)^{2s}}{\hat{s}}, \tag{9.49}$$

$$\rho^{[0]}_{II} = \frac{(-)^{2I}}{\hat{I}}, \tag{9.50}$$

we obtain from eq.(9.48)

$$P_{\alpha\beta}(k_z,\hat{k},\rho's) = \sum \frac{\hat{\Lambda}}{\hat{C}^4\ \hat{s}_a\ \hat{s}_b\ \hat{I}_A\ \hat{I}_B}$$

$$\times \begin{bmatrix} s_a & s_a & 0 \\ \ell_a & \ell'_a & \Lambda \\ j_a & j'_a & \Lambda \end{bmatrix} \begin{bmatrix} s_b & s_b & 0 \\ \ell_b & \ell'_b & \Lambda \\ j_b & j'_b & \Lambda \end{bmatrix} \begin{bmatrix} j_a & j'_a & \Lambda \\ I_A & I_A & 0 \\ C & C & \Lambda \end{bmatrix} \begin{bmatrix} j_b & j'_b & \Lambda \\ I_B & I_B & 0 \\ C & C & \Lambda \end{bmatrix}$$

$$\times \tilde{\mathcal{U}}_{\alpha\gamma}(C)\ \tilde{\mathcal{V}}_{\gamma\beta}(C)\ \mathcal{V}_{\beta\gamma}(C)\ \mathcal{U}_{\gamma\alpha}(C)\ [\rho^{[\Lambda]}_{\ell_a\ell'_a}(k_z)\ \rho^{[\Lambda]}_{\ell_b\ell'_b}(\hat{k})]^{[0]}, \tag{9.51}$$

with

$$\left[\rho^{[\Lambda]}_{\ell_a\ell'_a}(k_z)\ \rho^{[\Lambda]}_{\ell_b\ell'_b}(\hat{k})\right]^{[0]} = (\Lambda\,\Lambda\,00|00)\ \rho^{[\Lambda]}_{\ell_a\ell'_a0}(k_z)\ \rho^{[\Lambda]}_{\ell_b\ell_b0}(\hat{k})$$

$$= \frac{1}{4\pi}\frac{1}{\hat{\Lambda}}\,[\ell_a|\ell'_a|\Lambda]\ [\ell_b|\ell'_b|\Lambda]\ P_\Lambda(\cos\theta). \tag{9.52}$$

As the final case we compute the probability for the reaction again in the absence of polarizations but when the outgoing reaction products are not detected (inclusive reaction, or total cross section). In that case in addition to the above assignement of quantum numbers we have $\Lambda = 0$. The beam density matrices now the reduce to the simple forms

$$\rho^{[0]}_{\ell_a\ell'_a}(k_z) = \frac{1}{4\pi}\frac{1}{\hat{\ell}_a}\,\delta_{\ell_a\ell'_a}, \tag{9.53}$$

$$\int d^2\hat{k}\ \rho^{[0]}_{\ell_b\ell'_b}(\hat{k}) = \frac{1}{\hat{\ell}_b}\,\delta_{\ell_b\ell'_b}. \tag{9.54}$$

Inserting these values in eq.(9.51) we obtain

$$P_{\alpha\beta} = \frac{1}{4\pi} \sum \frac{1}{\left(\hat{C}\ \hat{s}_a\ \hat{s}_b\ \hat{\ell}_a\ \hat{\ell}_b I_A I_B\right)^2} \tilde{\mathcal{U}}_{\alpha\gamma}(C)\ \tilde{\mathcal{V}}_{\gamma\beta}(C)\ \mathcal{V}_{\beta\gamma}(C)\ \mathcal{U}_{\gamma\alpha}(C). \tag{9.55}$$

Chapter 10

SUMMARY AND FORMULARY

In this Chapter we give a complete summary of the tools necessary for carrying out calculations with the Invariant Graph method.

Rules for drawing graphs

Any quantity to be computed is first written in terms of an invariant matrix element. This invariant matrix element and all the recoupling transformations needed for its evaluation are represented by a graph.

In order to draw and evaluate this graph, one obeys the following rules :

(1) The diagram starts on the left with lines drawn from top to bottom in the order in which the tensorial quantities appear in the initial matrix element read from left to right. All numerical factors in front of the invariant matrix element are written on the left side of the graph.

(2) Only two adjacent lines or two adjacent already coupled groups of lines can be coupled. Couplings are indicated by brackets.

(3) The lines undergo a number of recouplings leading to end-boxes which

represent the elementary invariant matrix elements. These recouplings are drawn so as to respect the rules :

- Only lines which are coupled can be recoupled. A group of lines coupled to a given angular momentum which remains intact in a recoupling behaves like a single line with that same angular momentum.
- Lines can cross only within a recoupling box.

(4) All quantum numbers arising in the graph as a result of recoupling imply summation over their values.

This way, one first will draw a recoupling diagram. Then one will try to simplify it by choosing a different order of recoupling, until one is satisfied that no further improvements are possible.

Only at this time is it necessary to supply the graph with labels for the new quantum numbers arising from recouplings. This is done begining at the right hand side of the graph writing directly the particular values imposed by the final invariant products and end-boxes.

Now the final algebraic expression can be directly read off the graph. The algebraic result is obtained by writing the product of all the values associated with the graph symbols of the diagram, including all numerical factors. This expression is complete : no additional factors or phases have to be supplied.

Graph elements

<u>Coupling of 2 tensors</u>, $[\psi^{[i]}\psi^{[j]}]^{[K]}$, fig.10.1.

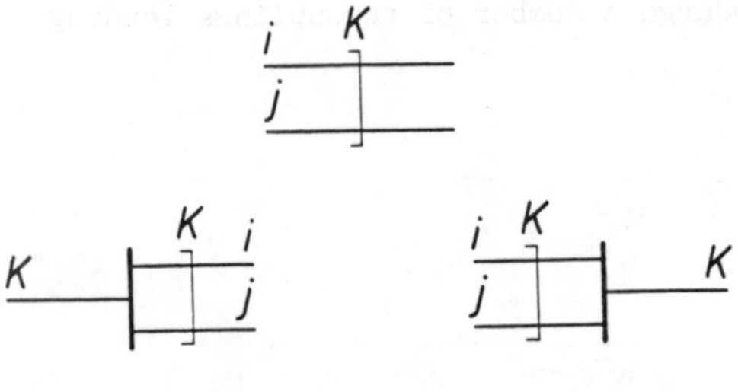

10.1

Permutation of 2 tensors, $(-)^{i+j-K}$, fig. 10.2

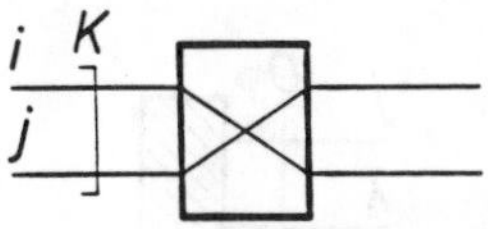

10.2

Invariant triple product, $[\psi^{[i]}\psi^{[j]}\psi^{[k]}]^{[0]}$, fig.10.3

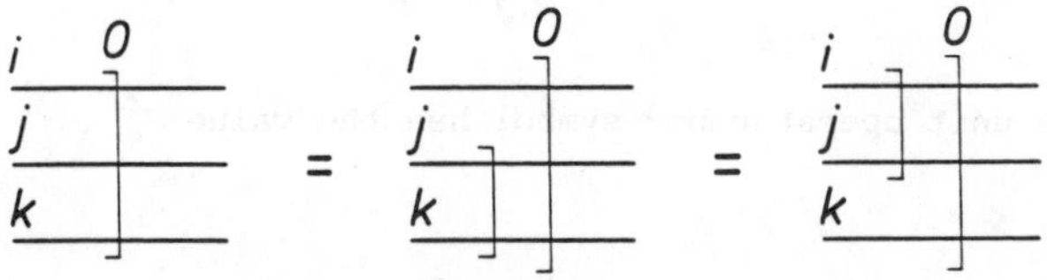

10.3

Invariant overlap of a state vector, $\left[\psi^{[j]}\middle|\psi^{[j]}\right] = \hat{j}$, fig.10.4

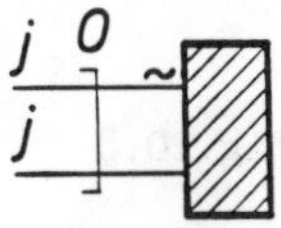

10.4

Invariant overlap of an amplitude vector, $\left[W^{[j]}\middle|W^{[j]}\right] = \frac{1}{\hat{j}}$, fig.10.5

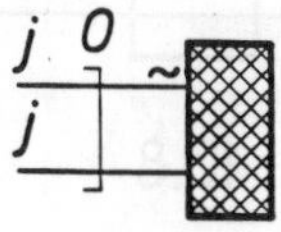

10.5

<u>Invariant matrix element</u>, $\left[\psi^{[j]}\middle|\Omega^{[\lambda]}\middle|\varphi^{[k]}\right]$, fig.10.6

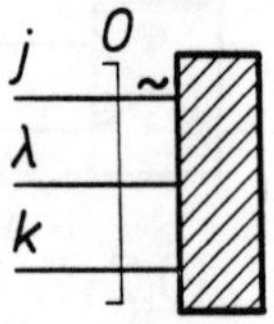

10.6

<u>Insertion of the unit operator</u>, $1 = \sum_{\alpha,j} \hat{j} \left[\psi_{\alpha}^{[j]}\middle|\psi_{\alpha}^{[j]}\right]$, fig.10.7

In a graph the unit operator box symbol has the value $\hat{j}$.

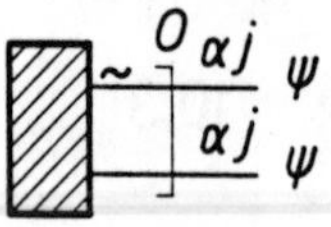

10.7

<u>Recoupling of 4 tensors</u>, $\begin{bmatrix} a & b & e \\ c & d & f \\ g & h & i \end{bmatrix}$, fig.10.8

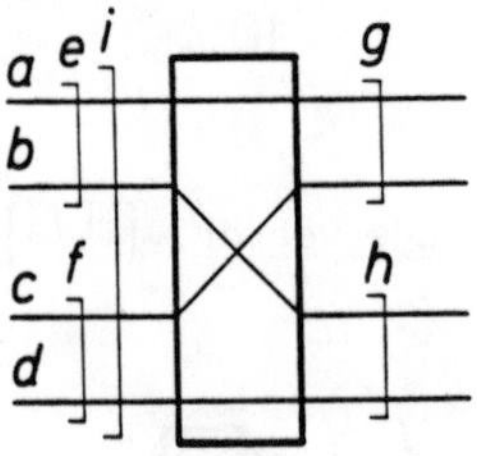

10.8

$$\begin{bmatrix} a & b & e \\ c & d & f \\ g & h & i \end{bmatrix} = \hat{e}\,\hat{f}\,\hat{g}\,\hat{h} \begin{Bmatrix} a & b & e \\ c & d & f \\ g & h & i \end{Bmatrix}.$$

Recoupling of 3 tensors, $\begin{bmatrix} 0 & b & b \\ c & d & f \\ c & h & i \end{bmatrix}$, $\begin{bmatrix} a & 0 & a \\ c & d & f \\ g & d & i \end{bmatrix}$, etc..., fig.10.9

In terms of the Wigner 6-j coefficients, see Table 10.1.

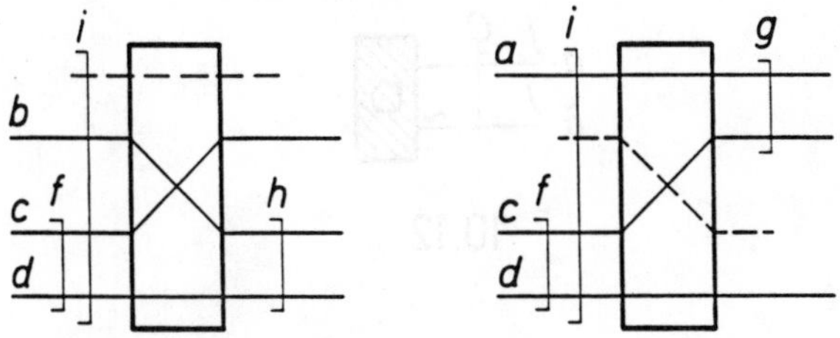

10.9

One-particle CFP, $(j^{n-1}\beta K\ ;\ j|\}j^n\alpha I)$, fig.10.10

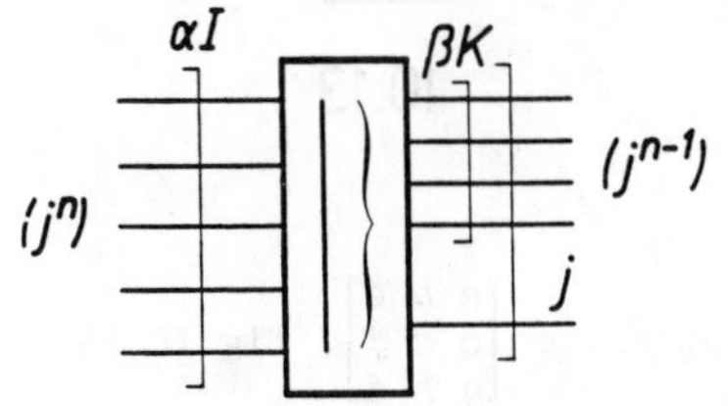

10.10

Two-particle CFP, $(j^{n-2}\beta K\ ;\ j^2L||\}j^n\alpha I)$, fig.10.11

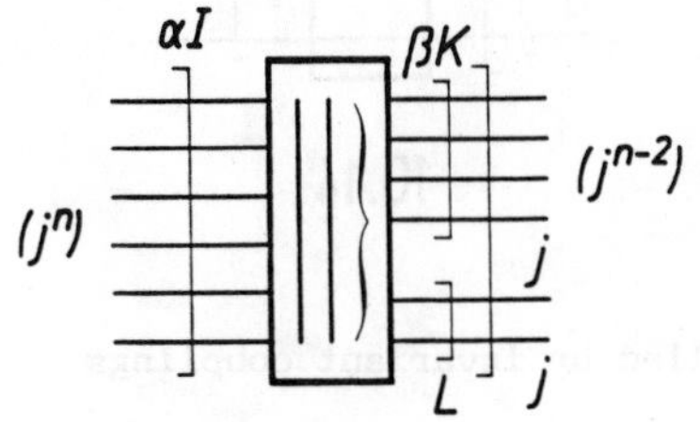

10.11

Contraction of Fock operators,

$\langle 0|\, [a^{[j]}\, \tilde{a}^{[j']}]^{[0]}\, |0\rangle = \epsilon\, \hat{j}\, \delta_{jj'}$, where $\epsilon = \pm 1$ for bosons and for fermions respectively, fig 10.12

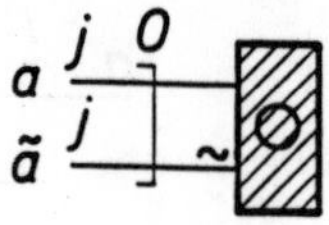

10.12

Crossing of Fock operators, $\epsilon(-)^{i+j-K}$, fig.10.13

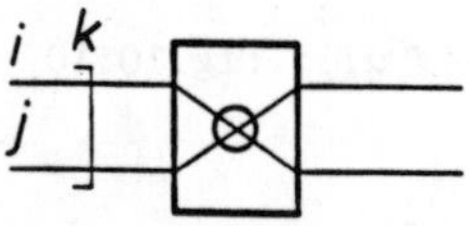

10.13

Recoupling of Fock operators, $\epsilon \begin{bmatrix} a & b & e \\ c & d & f \\ g & h & i \end{bmatrix}$, fig.10.14

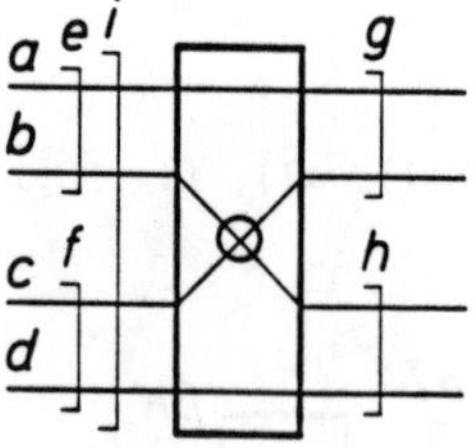

10.14

Useful relations fulfilled by invariant couplings

All the coupling schemes which respect the order of the lines within an invariant triple product are equivalent, fig.10.3 and 10.15

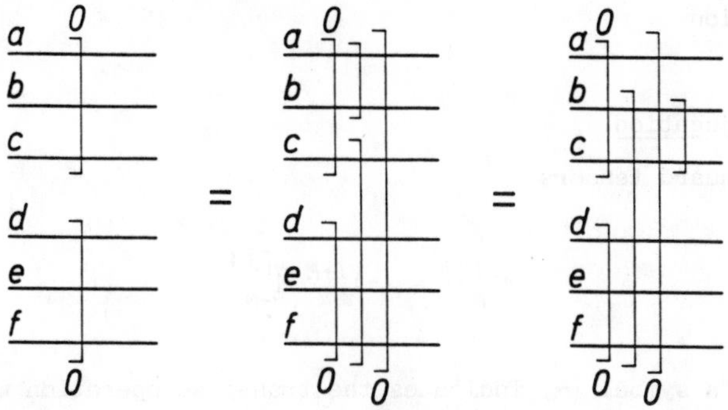

10.15

For lines representing commuting quantities an invariant can be inserted at any position between the graph lines, fig.10.16

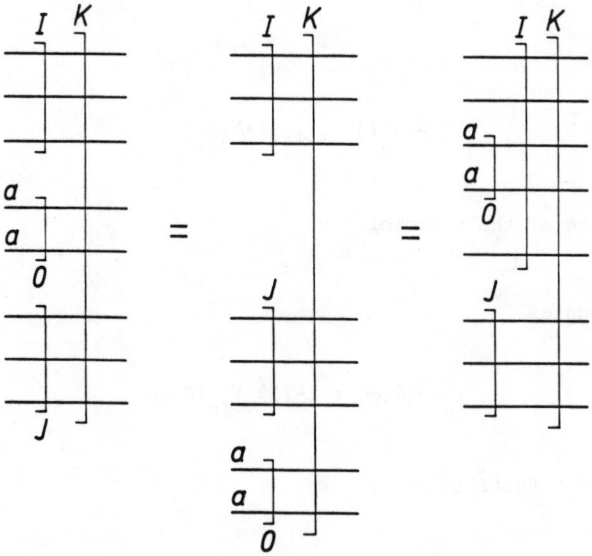

10.16

Phase convention

<u>Hermitian conjugation</u>

For contrastandard tensors

$$A^{[I]\dagger}_{M} = (-)^{I+M}\, \tilde{A}^{[I]}_{-M}$$

where the tilda symbol (~) indicates the transpose operation when A is a spinor or matrix.

In Fock space

$$a^{[I]\dagger}_{M} = (-)^{I+M}\, \tilde{a}^{[I]}_{-M}$$

where $a^{[I]}_{M}$ is an annihilation operator and where the ~ symbol denotes that $\tilde{a}^{[I]}_{M}$ is a creation operator.

<u>Time reversal</u>

$$T(A^{[I]}_{M}) = (-)^{I+M} A^{[I]}_{-M}$$

Note the absence of the ~ symbol.

<u>Spherical harmonics</u>

$$Y^{[\ell]}_{m}(\theta,\varphi) = (-i)^{\ell}\, Y_{\ell m}(\theta,\varphi)$$

where the $Y_{\ell m}(\theta\varphi)$ fulfill

$$Y^{*}_{\ell m}(\theta,\varphi) = (-)^{m}\, Y_{\ell -m}(\theta,\varphi)$$

<u>Spin 1 or unit vector</u>

$$e^{[1]}_{1} = i \frac{1}{\sqrt{2}} (\vec{e}_x + i\vec{e}_y)$$

$$e^{[1]}_{0} = - i\vec{e}_z$$

$$e^{[1]}_{-1} = - i \frac{1}{\sqrt{2}} (\vec{e}_x - i\vec{e}_y)$$

Vector amplitudes

$$V^{[1]}_{1} = i \frac{1}{\sqrt{2}} (V_x + iV_y)$$

$$V^{[1]}_{0} = -iV_z$$

$$V^{[1]}_{-1} = -i \frac{1}{\sqrt{2}} (V_x - iV_y)$$

Invariant forms

Wave functions

$$\psi_I = \hat{I} \, [W^{[I]}\psi^{[I]}]^{[0]}$$

Operators

$$\Omega_\lambda = \hat{\lambda} \, [\omega^{[\lambda]}\Omega^{[\lambda]}]^{[0]}$$

For scalar operators, $\lambda=0$, $\omega^{[0]}=1$.

Cartesian vectors

$$\vec{A} = \hat{1} \, [A^{[1]}e^{[1]}]^{[0]}$$

Scalar product

$$S = \vec{A}\cdot\vec{B} = \hat{1}\ [A^{[1]}B^{[1]}]^{[0]}$$

Vector product

$$\vec{V} = \vec{A} \times \vec{B} = \hat{1}\ \sqrt{2}\ [e^{[1]}A^{[1]}B^{[1]}]^{[0]}$$

Gradient

$$\vec{\nabla} = \hat{1}\ [e^{[1]}\nabla^{[1]}]^{[0]}$$

Divergence

$$\vec{\nabla}\cdot\vec{A} = \hat{1}\ [\nabla^{[1]}A^{[1]}]^{[0]}$$

Curl

$$\vec{\nabla} \times \vec{A} = \hat{1}\ \sqrt{2}\ [e^{[1]}\nabla^{[1]}A^{[1]}]^{[0]}$$

Density matrices

$$\rho^{[L]}_{\ell\ell' M} = \left[W^{[\ell]}\ \tilde{W}^{[\ell']}\right]^{[L]}_{M}$$

Matrix elements

Norm matrix element

$$\langle\psi_I|\psi_I\rangle = \hat{I}^2\left[[W^{[I]}\psi^{[I]}]^{[0]}\middle|[W^{[I]}\psi^{[I]}]^{[0]}\right]$$

$$= \frac{1}{\hat{I}}\ [\psi^{[I]}|\psi^{[I]}] = 1$$

for a polarized state, $\left|W^{[I]}_{M'}\right| = \delta_{M'M}$

$$\langle\psi_I|\psi_I\rangle = \langle\psi_M^{[I]}|\psi_M^{[I]}\rangle = \frac{1}{\tilde{I}}[\psi^{[I]}|\psi^{[I]}]$$

Operator matrix elements

Wigner-Eckart Theorem

$$\int \psi_{fM}^{[I]\dagger}\,\Omega_{\mu}^{[\lambda]}\,\psi_{iM'}^{[J]} = (-)^{I+M}(-)^{I-\lambda+J}\begin{pmatrix} I & \lambda & J \\ -M & \mu & M' \end{pmatrix}\left[\psi_f^{[I]}\right|\Omega^{[\lambda]}\left|\psi_i^{[J]}\right]$$

General case

$$\left\langle\psi_{fI}|\Omega_{\lambda}|\psi_{ij}\right\rangle = \hat{I}\,\hat{\lambda}\,\hat{J}\left[\left[W_f^{[I]}\psi_f^{[I]}\right]^{[0]}\left|[\omega^{[\lambda]}]\Omega^{[\lambda]}]^{[0]}\right|\left[W_i^{[J]}\psi_i^{[J]}\right]^{[0]}\right]$$

$$= \left[\tilde{W}_f^{[I]}\omega^{[\lambda]}W_i^{[J]}\right]^{[0]}\left[\psi_f^{[I]}|\Omega^{[\lambda]}|\psi_i^{[J]}\right]$$

Scalar operator

$$\left\langle\psi_{fI}\middle|\Omega_0\middle|\psi_{iI}\right\rangle = \frac{1}{\hat{I}}\left[\psi_f^{[I]}\middle|\Omega^{[0]}\middle|\psi_i^{[I]}\right]$$

and for polarized states, $\left|W_{iM'}^{[I]}\right| = \left|W_{fM'}^{[I]}\right| = \delta_{M'M}$

$$\left\langle\psi_{fM}^{[I]}|\Omega_0|\psi_{iM}^{[I]}\right\rangle = \frac{1}{\hat{I}}\left[\psi_f^{[I]}\middle|\Omega^{[0]}\middle|\psi_i^{[I]}\right]$$

Table of most common invariant matrix elements

Unit operator

$$[\psi^{[I]}|1^{[0]}|\psi^{[I]}] = [\psi^{[I]}|\psi^{[I]}] = \sqrt{2I+1} = \hat{I}$$

Spin operator

$$[\chi^{[\frac{1}{2}]}|\sigma^{[1]}|\chi^{[\frac{1}{2}]}] = [s|\sigma^{[1]}|s] = i\sqrt{6}$$

Angular momentum operator

$$[\psi^{[I]}|L^{[1]}|\psi^{[I]}] = i\,\hat{I}\,\sqrt{I(I+1)}$$

Spherical harmonic operator

$$[Y^{[\ell_1]}|Y^{[\ell_2]}|Y^{[\ell_3]}] = [\ell_1|\ell_2|\ell_3] = (-)^{(\ell_1+\ell_2+\ell_3)/2}\,\frac{\hat{\ell}_1\hat{\ell}_2\hat{\ell}_3}{\sqrt{4\pi}}\begin{pmatrix}\ell_1 & \ell_2 & \ell_3\\ 0 & 0 & 0\end{pmatrix}$$

Momentum and position operators

Denoting an orbital wave function of a complete orthonormal set

$$\varphi^{[\ell]}_{\alpha m}(\vec{r}) = F_{\alpha\ell}(r)\,Y^{[\ell]}_m(\hat{r})$$

and the operators of the radius vectors in position and momentum space

$$\vec{r} \Leftrightarrow r^{[1]}_m = \sqrt{\frac{4\pi}{3}}\;r\;Y^{[1]}_m(\hat{r})$$

$$\vec{p} \Leftrightarrow p^{[1]}_m = -i\;\nabla^{[1]}_m$$

the invariant matrix elements are

$$\left[\varphi^{[\ell]}_{\alpha}\middle|r^{[1]}\middle|\varphi^{[\ell-1]}_{\beta}\right] = \sqrt{\ell}\int r^2dr\;F_{\alpha\ell}(r)\;r\;F_{\beta\ell-1}(r)$$

$$\left[\varphi^{[\ell]}_{\alpha}\middle|r^{[1]}\middle|\varphi^{[\ell+1]}_{\beta}\right] = \sqrt{\ell+1}\int r^2dr\;F_{\alpha\ell}(r)\;r\;F_{\beta\ell+1}(r)$$

and

$$\left[\varphi_{\alpha}^{[\ell]}\middle|\nabla^{[1]}\middle|\varphi_{\beta}^{[\ell-1]}\right] = i\left[\varphi_{\alpha}^{[\ell]}\middle|p^{[1]}\middle|\varphi_{\beta}^{[\ell-1]}\right]$$

$$= \sqrt{\ell}\int r^2dr\, F_{\alpha\ell}(r)\left(\frac{\ell-1}{r} - \frac{\partial}{\partial r}\right)F_{\beta\ell-1}(r)$$

$$\left[\varphi_{\alpha}^{[\ell]}\middle|\nabla^{[1]}\middle|\varphi_{\beta}^{[\ell+1]}\right] = i\left[\varphi_{\alpha}^{[\ell]}\middle|p^{[1]}\middle|\varphi_{\beta}^{[\ell+1]}\right]$$

$$= \sqrt{\ell+1}\int r^2dr\, F_{\alpha\ell}(r)\left(\frac{\ell+2}{r} - \frac{\partial}{\partial r}\right)F_{\beta\ell+1}(r)$$

If the radial functions $F_{\alpha\ell}(r)$ are normalized spherical Bessel functions

$$F_{p\ell}(r) = \sqrt{\frac{2}{\pi}}\, j_{\ell}(pr)$$

the relations

$$j_{\ell+1}(x) = \left(\frac{\ell}{x} - \frac{\partial}{\partial x}\right) j_{\ell}(x)$$

$$j_{\ell-1}(x) = \left(\frac{\ell+1}{x} - \frac{\partial}{\partial x}\right) j_{\ell}(x)$$

and the orthonormality

$$\frac{2}{\pi}\int r^2dr\, j_{\ell}(pr)j_{\ell}(p'r) = \frac{\delta(p-p')}{p^2}$$

yield for the invariant matrix element of the gradient operator

$$\left[\varphi_{p}^{[\lambda]}\middle|\nabla^{[1]}\middle|\varphi_{q}^{[\ell]}\right] = \frac{\delta(p-q)}{p^2}\, p \times \begin{cases} \sqrt{\lambda+1} & if\ \ell = \lambda+1 \\ -\sqrt{\lambda} & if\ \ell = \lambda-1 \end{cases}$$

Tables for recoupling coefficients with zeroes

Recoupling coefficients with one zero can be expressed in terms of 6-j coefficients, table 10.1. Actually it is recommended not to do this replacement in the final formal expressions, because of the phases and complications it introduces.

On the other hand recoupling coefficients with two and three zeroes, Tables 10.2 and 10.3 respectively, have simple expressions and one should use these values to simplify the final algebraic form of the results.

Table 10.1

Recoupling coefficients with one zero

$$\begin{bmatrix} e & f & b \\ c & d & b \\ a & a & 0 \end{bmatrix} = (-)^{f+c+a+b}\hat{a}\hat{b}\begin{Bmatrix} e & f & b \\ d & c & a \end{Bmatrix}$$

$$\begin{bmatrix} 0 & a & a \\ b & d & c \\ b & f & e \end{bmatrix} = \begin{bmatrix} d & b & c \\ a & 0 & a \\ f & b & e \end{bmatrix} = (-)^{f+c+a+b}\hat{f}\hat{c}\begin{Bmatrix} e & f & b \\ d & c & a \end{Bmatrix}$$

$$\begin{bmatrix} b & d & c \\ 0 & a & a \\ b & f & e \end{bmatrix} = \begin{bmatrix} a & 0 & a \\ d & b & c \\ f & b & e \end{bmatrix} = (-)^{e+d+a+b}\hat{f}\hat{c}\begin{Bmatrix} e & f & b \\ d & c & a \end{Bmatrix}$$

$$\begin{bmatrix} a & a & 0 \\ e & f & b \\ c & d & b \end{bmatrix} = \begin{bmatrix} a & e & c \\ a & f & d \\ 0 & b & b \end{bmatrix} = (-)^{f+c+a+b}\frac{\hat{c}\hat{d}}{\hat{a}}\begin{Bmatrix} e & f & b \\ d & c & a \end{Bmatrix}$$

$$\begin{bmatrix} e & f & b \\ a & a & 0 \\ c & d & b \end{bmatrix} = \begin{bmatrix} e & a & c \\ f & a & d \\ b & 0 & b \end{bmatrix} = (-)^{e+d+a+b}\frac{\hat{c}\hat{d}}{\hat{a}}\begin{Bmatrix} e & f & b \\ d & c & a \end{Bmatrix}$$

Table 10.2

Recoupling coefficients with two zeroes

$$\begin{bmatrix} e & 0 & e \\ 0 & f & f \\ e & f & b \end{bmatrix} = \begin{bmatrix} e & 0 & e \\ b & f & e \\ f & f & 0 \end{bmatrix} = \begin{bmatrix} e & b & f \\ 0 & f & f \\ e & e & 0 \end{bmatrix} = 1$$

$$\begin{bmatrix} 0 & e & e \\ f & 0 & f \\ f & e & b \end{bmatrix} = \begin{bmatrix} 0 & e & e \\ f & b & e \\ f & f & 0 \end{bmatrix} = \begin{bmatrix} b & e & f \\ f & 0 & f \\ e & e & 0 \end{bmatrix} = (-)^{e+f-b}$$

$$\begin{bmatrix} e & e & 0 \\ e & b & f \\ 0 & f & f \end{bmatrix} = \begin{bmatrix} b & e & f \\ e & e & 0 \\ f & 0 & f \end{bmatrix} = \frac{1}{\hat{e}\hat{e}}$$

$$\begin{bmatrix} e & e & 0 \\ b & e & f \\ f & 0 & f \end{bmatrix} = \begin{bmatrix} e & b & f \\ e & e & 0 \\ 0 & f & f \end{bmatrix} = \frac{(-)^{e+f-b}}{\hat{e}\hat{e}}$$

$$\begin{bmatrix} 0 & e & e \\ f & f & 0 \\ f & b & e \end{bmatrix} = \begin{bmatrix} 0 & f & f \\ e & f & b \\ e & 0 & e \end{bmatrix} = \begin{bmatrix} e & e & 0 \\ f & 0 & f \\ b & e & f \end{bmatrix} = \begin{bmatrix} e & f & b \\ e & 0 & e \\ 0 & f & f \end{bmatrix} = \frac{\hat{b}}{\hat{e}\hat{f}}$$

$$\begin{bmatrix} e & 0 & e \\ f & f & 0 \\ b & f & e \end{bmatrix} = \begin{bmatrix} e & f & b \\ 0 & f & f \\ e & 0 & e \end{bmatrix} = \begin{bmatrix} e & 0 & e \\ e & f & b \\ 0 & f & f \end{bmatrix} = \begin{bmatrix} e & e & 0 \\ 0 & f & f \\ e & b & f \end{bmatrix} = (-)^{e+f-b}\frac{\hat{b}}{\hat{e}\hat{f}}$$

Table 10.3

Recoupling coefficients with three zeroes

$$\begin{bmatrix} e & f & b \\ 0 & 0 & 0 \\ e & f & b \end{bmatrix} = \begin{bmatrix} e & 0 & e \\ f & 0 & f \\ b & 0 & b \end{bmatrix} = \begin{bmatrix} 0 & 0 & 0 \\ e & f & b \\ e & f & b \end{bmatrix} = \begin{bmatrix} 0 & e & e \\ 0 & f & f \\ 0 & b & b \end{bmatrix} = 1$$

$$\begin{bmatrix} e & f & b \\ e & f & b \\ 0 & 0 & 0 \end{bmatrix} = \begin{bmatrix} e & e & 0 \\ f & f & 0 \\ b & b & 0 \end{bmatrix} = \frac{\hat{b}}{\hat{e}\hat{f}}$$

$$\begin{bmatrix} e & 0 & e \\ e & e & 0 \\ 0 & e & e \end{bmatrix} = \begin{bmatrix} e & e & 0 \\ 0 & e & e \\ e & 0 & e \end{bmatrix} = (-)^{2e}\frac{1}{\hat{e}\hat{e}}$$

$$\begin{bmatrix} e & e & 0 \\ e & 0 & e \\ 0 & e & e \end{bmatrix} = \begin{bmatrix} 0 & e & e \\ e & e & 0 \\ e & 0 & e \end{bmatrix} = \frac{1}{\hat{e}\hat{e}}$$

INDEX

INDEX